Lecture Notes in Control and Information Sciences

Edited by A. V. Balakrishnan and M. Thoma

Vol. 22: Optimization Techniques
Proceedings of the 9th IFIP Conference on Optimization Techniques,
Warsaw, September 4–8, 1979
Part 1
Edited by K. Iracki, K. Malanowski, S. Walukiewicz
XVI, 569 pages. 1980

Vol. 23: Optimization Techniques
Proceedings of the 9th IFIP Conference on Optimization Techniques,
Warsaw, September 4-8, 1979
Part 2
Edited by K. Iracki, K. Malanowski, S. Walukiewicz
XV, 621 pages. 1980

Vol. 24: Methods and Applications in Adaptive Control
Proceedings of an International Symposium
Bochum, 1980
Edited by H. Unbehauen
VI, 309 pages. 1980

Vol. 25: Stochastic Differential Systems – Filtering and Control
Proceedings of the IFIP-WG7/1 Working Conference
Vilnius, Lithuania, USSR, Aug. 28 – Sept. 2, 1978
Edited by B. Grigelionis
X, 362 pages. 1980

Vol. 26: D. L. Iglehart, G. S. Shedler
Regenerative Simulation of Response Times in Networks of Queues
XII, 204 pages. 1980

Vol. 27: D. H. Jacobson, D. H. Martin, M. Pachter, T. Geveci
Extensions of Linear-Quadratic Control Theory
XI, 288 pages. 1980

Vol. 28: Analysis and Optimization of Systems
Proceedings of the Fourth International Conference on Analysis and Optimization of Systems
Versailles, December 16–19, 1980
Edited by A. Bensoussan and J. L. Lions
XIV, 999 pages. 1980

Vol. 29: M. Vidyasagar,
Input-Output Analysis of Large-Scale Interconnected Systems – Decomposition, Well-Posedness and Stability
VI, 221 pages. 1981

Vol. 30: Optimization and Optimal Control
Proceedings of a Conference Held at Oberwolfach, March 16–22, 1980
Edited by A. Auslender, W. Oettli, and J. Stoer
VIII, 254 pages. 1981

Vol. 31: Berc Rustem
Projection Methods in Constrained Optimisation and Applications to Optimal Policy Decisions
XV, 315 pages. 1981

Vol. 32: Tsuyoshi Matsuo,
Realization Theory of Continuous-Time Dynamical Systems
VI, 329 pages, 1981

Vol. 33: Peter Dransfield
Hydraulic Control Systems – Design and Analysis of Their Dynamics
VII, 227 pages, 1981

Vol. 34: H.W. Knobloch
Higher Order Necessary Conditions in Optimal Control Theory
V, 173 pages, 1981

Vol. 35: Global Modelling
Proceedings of the IFIP-WG 7/1 Working Conference Dubrovnik, Yugoslavia,
Sept. 1–5, 1980
Edited by S. Krčevinac
VIII, 232 pages, 1981

Vol. 36: Stochastic Differential Systems
Proceedings of the 3rd IFIP-WG 7/1 Working Conference
Visegrád, Hungary, Sept. 15–20, 1980
Edited by M. Arató, D. Vermes, A.V. Balakrishnan
VI, 238 pages, 1981

Vol. 37: Rüdiger Schmidt
Advances in Nonlinear Parameter Optimization
VI, 159 pages, 1982

Vol. 38: System Modeling and Optimization
Proceedings of the 10th IFIP Conference
New York City, USA, Aug. 31 – Sept. 4, 1981
Edited by R.F. Drenick and F. Kozin
XI, 894 pages. 1982

Vol. 39: Feedback Control of Linear and Nonlinear Systems
Proceedings of the Joint Workshop on Feedback and Synthesis of Linear and Nonlinear Systems
Bielefeld/Rom
XIII, 284 pages. 1982

Vol. 40: Y.S. Hung, A.G.J. MacFarlane
Multivariable Feedback: A Quasi-Classical Approach
X, 182 pages. 1982

Vol. 41: M. Gössel
Nonlinear Time-Discrete Systems – A General Approach by Nonlinear Superposition
VIII, 112 pages. 1982

Vol. 42: Advances in Filtering and Optimal Stochastic Control
Proceedings of the IFIP-WG 7/1 Working Conference
Cocoyoc, Mexico, February 1–6, 1982
VIII, 391 pages. 1982

For information about Vols. 1–21 please contact your bookseller or Springer-Verlag.

Lecture Notes in Control and Information Sciences

Edited by M. Thoma and A. Wyner

93

K. Malanowski

Stability of Solutions to Convex Problems of Optimization

Springer-Verlag
Berlin Heidelberg GmbH

Author
K. Malanowski
Systems Research Institute
of the Polish Academy of Sciences
ul. Newelska 6
01-447 Warszawa
Poland

Library of Congress Cataloging in Publication Data
Malanowski, Kazimierz.
Stability of solutions to convex problems of optimization.
(Lecture notes in control and information sciences ; 93)
Bibliography: p.
1. Control theory.
2. Stability.
3. Mathematical optimization.
4. Convex functions.
I. Title.
II. Series.
QA402.3.M329 1987 629.8'312 87-4766

ISBN 978-3-540-17589-6 ISBN 978-3-540-47707-5 (eBook)
DOI 10.1007/978-3-540-47707-5

Originally published by Springer-Verlag Berlin Heidelberg New York in 1987.

2161/3020-543210

INTRODUCTION

Problems of stability with respect to data of optimization problems can be divided into two groups: global stability, which is investigated over the whole domain of values of parameters, and differential stability, called also sensitivity, where only local changes of parameters are investigated.

In each case stability is understood either in the sense of solutions to optimization problems or in the sense of the so-called optimal value function (called also value function, extremal value function or marginal function), which to every value of the parameter assigns the corresponding optimal value of the cost functional.

For more than ten years stability of optimization problems has been intensively studied and stability analysis is fairly well developed, although it is far from being complete.

The investigations concern stability of both mathematical programming problems in finite dimensional spaces and of optimization problems in functional spaces. These latter problems are studied either in abstract formulations or in more specialized forms among which optimal control problems play an important role.

As one could expect the results obtained for finite dimensional problems are more complete and constructive than those for general cases. It is also not surprising that stability properties of optimal value function have been obtained for much more general situations and under weaker assumptions than those for solutions of optimization problems.

Some aspects of stability problems are treated in a synthetic form in several monographs, which appeared in recent years [4, 9, 12, 17, 22].

In global stability analysis the notion of solution set plays an important role. The solution set is defined as the set of those feasible points at which the cost functional assumes its minimal value.

The properties of the solution set have been intensively studied using the general results of the theory of set-to-point mappings.

Numerous interesting results concerning semicontinuity, continuity and Lipschitz continuity of solution sets, as the function of parameters, and of the optimal value function have been obtained. A survey of the results in that area and an extensive bibliography can be found in [4].

Under stronger assumptions, where the solutions to the analysed optimal control problems are unique, the solution sets reduce to singletons and respective point-to-set mappings become usual functions.

Various specialized results concerning local and global properties of solutions to optimization problems and associated Lagrange multipliers as functions of parameters have been obtained. These results encompass conditions of continuity [32, 34], Lipschitz continuity [3, 10, 13, 24] and differentiability [5, 11, 15, 16, 30, 31, 45]. A comprehensive survey of the results concerning mostly mathematical programming problems can be found in [17], while those concerning optimal control are discussed in [12] and in references thereof.

If seems that most of the effort in the stability analysis has been devoted to differential properties of optimal value function, where the advanced technique of non-differentiable optimization has been applied. The results obtained concern mathematical programming problems [2, 8, 18, 46, 47], general problems of optimization [19, 21, 34] and some specialized problems like optimal control [20, 42]. For more details and literature concerning the subject the reader is referred to [9, 46, 47].

These notes are devoted to a rather narrow class of optimization problems, namely those which are strongly convex and subject to pointwise constraints, satisfying linear independence condition.

These two conditions yield uniqueness of both the solutions and the associated Lagrange multipliers.

It should be strongly stressed that the strict complementarity condition is not required.

Two classes of optimization problems are considered: mathematical programming in finite dimensional spaces and optimal control problems.

It is assumed that all the data of the problems depend on a finite dimensional vector parameter and that the set of admissible parameters is convex and open.

Part of the results could be easily extended to the case where parameters are elements of a functional space.

The analysis presented concerns both global and differential stability of the solutions and the associated Lagrange multipliers. Accordingly the material is organized into two parts devoted to these two classes of problems.

Our assumptions imply two important properties of the solutions and the associated Lagrange multipliers, namely that they are Lipschitz continuous and directionally (one-sidedly) differentiable functions of the vector parameter. The assumptions are not strong enough to provide continuous differentiability of these functions. If the assumption of strict complementarity is added, then continuous differentiability takes place (see [15]). However, the assumption that the strict complemetarity condition holds for all values of parameters of a mathematical program is very restrictive, while our assumptions hold for all values of parameters for a class of mathematical programming problems.

Since the case of differentiable solutions (under strict complementarity condition) is thoroughly discussed in the book by Fiacco [17], these notes, in the part concerning mathematical programming, can be treated as a supplement to that book. Since the purpose of these notes is to give a possibly complete picture of sensitivity problems for the discussed class of optimization problems, then an attempt is made to give proofs of all presented results. It is not intended to weaken the assumptions as far as possible. Quite opposite, many results, especially auxiliary ones, can be obtained under weaker assumptions. Some basic assumptions however seem to be indispensable. This refers in particular to strong convexity (which can be relaxed to the so called "second order sufficiency conditions" - [17]) and to pointwise linear independence of the binding constraints.

The first part of the notes is devoted to global stability and more precisely to Lipschitz continuity of the solutions and the associated Lagrange multipliers for mathematical programming and optimal control problems.

The central idea of this part is the abstract result due to W.W. Hager, concerning Lipschitz continuity of constrained proceses [24]. This result was applied by Hager himself to sensitivity analysis of convex programming problems and by A.L. Dontchev [12] to optimal control. All these results are presented here.

The Lipschitz continuity of solutions and Lagrange multipliers discussed in Part I is a necessary starting point to the analysis of differential stability performed in Part II.

The presented analysis is constructive in the sense that the method of finding the right-differential of solutions with respect to the parameter is given. The differential can be found as the solution to an auxiliary quadratic problem of optimization.

These results are derived using Lagrange formalism for initial optimization problems.

A crucial result of this part is the theorem due to K. Jittorntrum [30, 31], which gives the form of right-differentials of solutions and the associated Lagrange multipliers for convex programming problems that depend on a vector parameter. In the case where strict complementarity holds right-differentials become continuous ones.

The Jittorntrum's result is further developed by the author and used to obtain similar results for optimal control problems subject to pointwise constraints.

Note that in differential stability of solutions to constrained optimization problems in functional spaces very little has been done. One has to mention an alternative approach, not presented here, which is based on the concept of directional differentiability of the operation of projection onto a convex set in a Hilbert space. This idea was first introduced by A. Haraux [26, 27] and F. Mignot [43] who used it to stability analysis of variational inequalities. The approach has been further developed and applied to sensitivity analysis of optimal control problems by J. Sokołowski [48-51].

The author is fully aware that the presented material is far from being complete, even for the restricted class of problems discussed here. Therefore at the end of the notes some unsolved problems are formulated.

The enclosed bibliography is practically restricted to the positions directly quoted in the text.

Throughout the notes more or less standard notation is used. In particular:

$\langle\cdot,\cdot\rangle$ and $|\cdot|=\langle\cdot,\cdot\rangle^{\frac{1}{2}}$ denote the usual inner product and the associated norm in an Euclidean space,

$(\cdot,\cdot)$ and $||\cdot||=(\cdot,\cdot)^{1/2}$ denote the inner product and the norm in a Hilbert space.

In case of possible ambiguity the norms are supplied with appropriate subscripts.

If $f(\cdot,\cdot)$ is a properly regular function of two variables (u,h) then

$D_u f(u,h)$, $D^2_{uh} f(u,h)$ - denote continuous (Fréchet) derivatives of respective order, with respect to the appropriate variable.

For a properly regular function $\phi(\cdot)$ by

$$\delta^+_h\phi(h;g) \overset{\text{def}}{=} \lim_{\alpha\downarrow 0} \frac{1}{\alpha}\left[\phi(h+\alpha g)-\phi(h)\right]$$

is denoted the right-differential at the point h in the direction g, while

$d\phi(h;g)$ - denotes Gâteaux differential at the point h in the direction g.

$\mathcal{L}(X;Y)$ - denotes the space of linear continuous operators from a space X into Y.

Acknowledgement

My research in the field being the subject of these notes was initiated in the winter semester of 1981, when I was a visiting professor in the Institute of Applied Mathematics and Statistic of the University of Würzburg (Federal Republic of Germany). My stay there was supported by the Deutsche Forschungsgemeinschaft (DFG) which I gratefully acknowledge. I am very grateful to Professors H.W. Knobloch and J. Stoer as well as to Dr. B. Gollan from the University of Würzburg for stimulating discussions and suggestions. Especially it is my pleasure to forward my warm thanks to Professor J. Stoer, for very careful reading of two papers of mine, correcting errors and making important comments.

These notes would have never be written without my cooperation and constant discussions with Dr. Jan Sokołowski. In particular Dr. Sokołowski drew to my attention the results concerning differentiability of projection, which were the starting point to my own research in differential stability. I also discussed some problems with Dr. K. Kiwiel and Dr. J. Sosnowski. I extend to all of them my sincere thanks.

Warszawa, November 1985 K. Malanowski

CONTENTS

PART I

GLOBAL STABILITY

1. CONVEX PROGRAMMING PROBLEM

1.1. Problem Statement

Let $H \subset R^m$ be an open and convex set of vector parameters. Consider a family $\{P_h\}$ of the following convex programming problems depending on h

$$(P_h) \quad \left| \begin{array}{l} \text{find } u(h) \in R^n \text{ such that} \\ f(u(h),h) = \min\limits_{u \in \Phi_h} f(u,h), \end{array} \right. \tag{1.1.1}$$

where

$$\Phi_h = \{u \in R^n \mid \phi^i(u,h) \leq 0, \; i \in I\}, \tag{1.1.2}$$

$$I = \{1,2,\ldots,r\}.$$

We shall denote

$$\phi(u,h) = [\phi^1(u,h), \phi^2(u,h), \ldots, \phi^r(u,h)]^T.$$

Assume that the following conditions are satisfied:

(A1) for each $h \in H$ $f(\cdot,h)$ is twice continuously differentiable function of u. Moreover it is strongly convex, uniformly with respect to h, i.e. there exists a constant $\alpha > 0$ independent of h, such that

$$\langle v, D^2_{uu} f(u,h) v \rangle \geq \alpha |v|^2 \quad \forall u,v \in R^n, \; \forall h \in H \tag{1.1.3}$$

(A2) $f(\cdot,\cdot)$ and $D_u f(\circ,\cdot)$ are continuously differentiable functions on $R^n \times H$

(A3) for each $h \in H$ $\phi^i(\cdot,h)$, $i \in I$, are twice continuously differentiable and convex function of u

(A4) $\phi^i(\cdot,\cdot)$ and $D_u \phi^i(\circ,\cdot)$, $i \in I$, are continuously differentiable functions on $R^n \times H$

(A5) for each $h \in H$ the admissible set Φ_h is non-empty

$$\Phi_h \neq \emptyset \tag{1.1.4}$$

It is well known [6] that by assumptions (A1), (A3) and (A5) for each $h \in H$ Problem (P_h) has a unique solution $u(h)$, which can be characterized by the following variational inequality

$$\langle D_u f(u(h),h), u-u(h) \rangle \geq 0 \quad \forall u \in \Phi_h \tag{1.1.5}$$

Denote by

$$I_h = \{i \in I \mid \phi^i(u(h),h) = 0\} \qquad (1.1.6)$$

the set of the indices of all constraint functions binding at $u(h)$.

In addition to (i) through (v) we assume that at the points $u(h)$ the following constraint regularity condition holds

(A6) there exists a constant $\beta > 0$ such that

$$|D_u\phi^T_{I_h}(u(h),h)v| \geq \beta|v| \qquad (1.1.7)$$

for every $h \in H$ and for every v of appropriate dimension, where $D_u\phi^T_{I_h}(u(h),h)$ denotes the matrix whose columns are the gradients of all constraint functions binding at $u(h)$.

Note that (1.1.7) implies that for (P_h) the Slater's condition holds. Namely

$$\text{there exists } \hat{u}_h \in R^n \text{ such that } \phi^i(\hat{u}_h,h)<0,\ i \in I. \qquad (1.1.8)$$

Indeed let us find a vector $v_h \in R^n$ such that

$$D_u\phi_{I_h}(u(h),h)v_h = -k\mathbb{1}, \qquad (1.1.9)$$

where $\mathbb{1}$ is the unit vector of appropriate dimension and $k > 0$. By (1.1.7) there exists a vector v_h such that

$$|v_h| \leq \frac{k}{\beta}. \qquad (1.1.10)$$

Put $\hat{u}_h=u(h)+v_h$. Since by Taylor's formula

$$\phi^i(\hat{u}_h,h)=\phi^i(u(h),h)+\langle D_u\phi^i(u(h),h),v_h\rangle+o(v_h), \qquad (1.1.11)$$

then (1.1.9) and (1.1.10) imply that for k sufficiently small we have

$$\phi^i(\hat{u}_h,h) < 0 \qquad i \in I_h.$$

Similarly for $i \in I \setminus I_h$ we have $\phi^i(u(h),h)<0$ hence (1.10) and (1.11) yields (1.1.8) for $|v_h|$ sufficiently small. This completes the proof of (1.1.8). For Problem (P_h) let us define Lagrangian

$$L(.,.;.) : R^n \times R^p \times H \to R^1$$

$$L(u,\lambda;h) \overset{\text{def}}{=} f(u,h)+ \langle\lambda,\phi(u,h)\rangle, \qquad (1.1.12)$$

where $\lambda^T=(\lambda^1,\lambda^2,\dots,\lambda^r)$ is a Lagrange multiplier associated with the constraints.

It is well known[6, 17] that if condition (1.1.8) holds then there exists a (normal) Lagrange multiplier $\lambda(h)\geq 0$ such that the solution $u(h)$ to (P_h) is characterized by the following saddle-point condition

$$L(u(h),\lambda;h) \leq L(u(h),\lambda(h);h) \leq L(u,\lambda(h);h)$$

$$\forall u\in R^n,\ \forall\lambda\in R^r,\ \lambda^i \geq 0,\ i\in I. \tag{1.1.13}$$

Condition (1.1.13) is equivalent to the following Kuhn-Tucker conditions:

$$D_uL(u(h),\lambda(h);h)=D_uf(u(h),h)+D_u\phi^T(u(h),h)\lambda(h)=0, \tag{1.1.14}$$

$$\lambda^i(h)\phi^i(u(h),h) = 0\ , \tag{1.1.15}$$

$$\lambda^i(h) \geq 0\ , \qquad\qquad i\in I. \tag{1.1.16}$$

Note that (1.1.7) implies the uniqueness of $\lambda(h)$. Indeed by complementary slackness (1.1.15) we have

$$\lambda^i(h) = 0 \qquad \text{for} \qquad i\in I\setminus I_h \tag{1.1.17}$$

Let $\lambda_{I_h}(h)$ denote the subvector of $\lambda(h)$ containing all components $\lambda^i(h)$ for $i\in I_h$. Suppose that there are two different vectors $\lambda'_{I_h}(h)$ and $\lambda''_{I_h}(h)$ satisfying (1.1.14). Then by the uniqueness of $u(h)$ and by (1.1.17) we get

$$D_u\phi^T_{I_h}(u(h),h)\left[\lambda'_{I_h}(h)-\lambda''_{I_h}(h)\right] = 0$$

and by (1.1.7) $\lambda'_{I_h}(h)=\lambda''_{I_h}(h)$, which together with (1.1.17) show that Lagrange multipliers $\lambda(h)$ are defined uniquely.

1.2. Continuity of Solutions

Our purpose is to prove Lipschitz continuity, with respect to the parameter, of the solutions $u(h)$ to (P_h) as well as of the associated Lagrange multipliers $\lambda(h)$.

In this section we shall show continuity of these functions. To this end we shall need the following:

Lemma 1.1

For any $h\in H$ there exist constants $c(h)>0$ and $\rho(h)>0$ such that

$$|u(g)|,\ |\lambda(g)| \leq c(h) \qquad \forall g \in S(h,\rho(h)), \tag{1.2.1}$$

where

$$S(h,\rho) \overset{\text{def}}{=} \{g \,|\, |h-g| \leq \rho\} \tag{1.2.2}$$

is an open ball in R^m with the center at h and the radius ρ.

Proof.

By (1.1.8) together with (A1) and (A2) we find that for any $h \in H$ there exists a ball $S(h,\rho(h))$, such that

$$\phi^i(\hat{u}_h,g) < 0 \qquad \forall i \in I,\ \forall g \in S(h,\rho(h)) \tag{1.2.3}$$

i.e. $\hat{u}_h$ is an admissible element for (P_g), where $g \in S(h,\rho(h))$. Therefore

$$f(u_g,g) \leq f(\hat{u}_h,g). \tag{1.2.4}$$

Expanding $f(.,g)$ into Taylor's series at $\hat{u}_h$ and taking into account (1.1.3) we get

$$f(u_g,g) \geq f(\hat{u}_h,g)+(D_u f(\hat{u}_h,g),u_g-\hat{u}_h)+\alpha|u_g-\hat{u}_h|^2. \tag{1.2.5}$$

From (1.2.4) and (1.2.5) we obtain

$$|u_g-\hat{u}_h| \leq \frac{1}{\alpha}|D_u f(\hat{u}_h,g)| \qquad \forall g \in S(h,\rho(h)), \tag{1.2.6}$$

which together with (A2) imply the uniform boundedness of u_g on $S(h,\rho(h))$.

On the other hand from (1.1.7), (1.1.12) and (1.1.17) we get

$$|D_u f(u(g),g)| = |D_u\phi^T_{I_g}(u(g),g)\lambda_{I_g}(g)| \geq \beta|\lambda_{I_g}(g)| = \beta|\lambda(g)|$$

which by boundedness of $u(g)$ and by (A1) completes the proof of (1.2.1).

□

Corollary 1.1

For any compact set $\mathcal{H} \subset H$ there exists a constant $c > 0$ such that

$$|u(h)|,\ |\lambda(h)| \leq c \qquad \forall h \in \mathcal{H} \tag{1.2.7}$$

Proof.

For each $h \in H$ there exists a ball $S(h,\varepsilon(h))$ such that (1.2.1) holds. The set of these balls for all $h \in H$ constitutes an open covering of $\mathcal{H}$. From this covering we can extract a finite subcovering. This fact together with (1.2.1) imply (1.2.7).

Theorem 1.1.

Solutions $u(h)$ of (P_h) and associated Lagrange multipliers $\lambda(h)$ are continuous functions on H.

Proof.

Let $h \in H$ be an arbitrary point, and let $\{g\} \subset S(h,\varepsilon(h))$ be any sequence convergent to h

$$g \to h. \tag{1.2.8}$$

By Corollary 1.1 we can extract a subsequence $\{g'\} \subset \{g\}$ such that

$$u(g') \to \bar{u}, \quad \lambda(g') \to \bar{\lambda}. \tag{1.2.9}$$

By (1.1.13) we have

$$L(u(g'),\lambda;g') \leq L(u(g'),\lambda(g');g') \leq L(u,\lambda(g');g')$$

$$\forall u \in R^n, \quad \forall \lambda \in R^r, \quad \lambda^i \geq 0, \quad i \in I.$$

Hence (1.2.8), (1.2.9) together with (1.1.12) and (A1) through (A4) yield

$$L(\bar{u},\lambda,h) \leq L(\bar{u},\bar{\lambda},h) \leq L(u,\bar{\lambda},h) \tag{1.2.10}$$

$$\forall u \in R^n, \quad \forall \lambda \in R^r, \quad \lambda^i \geq 0, \quad i \in I.$$

On the other hand by (1.1.16) and (1.2.9) we have

$$\bar{\lambda}^i \geq 0 \qquad i \in I. \tag{1.2.11}$$

(1.2.10) together with (1.2.11) imply that $\bar{u}$ is a solution of (P_h) and $\bar{\lambda}$ is an associated multiplier. Since both the solution and the multipliers are unique we have $\bar{u}=u(h)$, $\bar{\lambda}=\lambda(h)$ and for the whole sequence $\{g\}$ we get

$$u(g) \to u(h),$$

$$\lambda(g) \to \lambda(h).$$

□

Continuity of $u(.)$ together with (A.3), (A4) and (1.1.15) imply

Corollary 1.2

For any $h \in H$ there exists a constant $\rho(h) > 0$ such that

$$I_g \subset I_h \qquad \forall g \in S(h,\rho(h)) \tag{1.2.12}$$

and

$$\lambda^i(g) = 0 \qquad i \in I \setminus I_h, \quad \forall g \in (S(h,\rho(h)) \tag{1.2.13}$$

1.3. Lipschitz Continuity of Solutions: Abstract Problem

We shall present the proof of Lipschitz continuity of (P_h) which is due to W.W. Hager [24]. This proof is based on an abstract result which is presented below. This is a general result, which can be applied not only to convex programming problem. Later on it will be used again.

Let H be a convex subset of a Banach space X (in the case of (P_h) we have $X=R^m$) and let Y be another Banach space.

On H there is defined a continuous mapping

$$k : H \to Y$$

and the mapping

$$I : H \to 2^{\{1,\dots,r\}},$$

where $2^{\{1,\dots,r\}}$ is the power set of $\{1,2,\dots,r\}$, having the following upper semicontinuity property

$$\text{if } \{h_k\} \subset H \text{ is any sequence such that } \lim_{k\to\infty} h_k = h \text{ and } J \subset I(h_k) \text{ for all } k, \text{ then } J \subset I(h). \tag{1.3.1}$$

In Problem (P_h) we shall put $Y=R^{n+r}$ and $k^T(h)=[u^T(h),\lambda^T(h)]$, while $I(h)=I_h$ will be the set of all constraints binding at $u(h)$.

For any ordered pair (g,h) by $\overline{[g,h]}$ is denoted the segment

$$\overline{[g,h]} = \{(1-s)g+sh \mid 0 \leqslant s \leqslant 1\}, \tag{1.3.2}$$

moreover we denote

$$\delta_s = ((1-s)g+sh)$$

Definition 1.1

The pair $(g,h) \in H \times H$ is called complatible if

$$I(g) = I(h) \quad\text{and}\quad I(\delta_s) \subset I(g) \quad \forall \delta_s \in \overline{[g,h]}. \tag{1.3.3}$$

Theorem 1.2

If there exist constants $\gamma < \infty$ and $\eta > 0$ such that

$$||k(h)-k(g)|| \leqslant \gamma\, ||h-g|| \tag{1.3.4}$$

for all pairs $(g,h) \in H \times H$ compatible and such that $||g-h|| \leqslant \eta$ then (1.3.4) holds for all pairs $(g,h) \in H \times H$.

Proof.(see [24])

Let $\#J$ denote the number of elements in a set $J \subset \{1,2,\ldots,r\}$. We define the following sets

$$T_q = \{(g,h) \in H \times H \mid \exists\, J \subset \{1,2,\ldots,r\} \text{ with } \#J \leqslant q \text{ and } I(\delta) \subset J \text{ for all } \delta \in \overline{[g,h]}\}. \tag{1.3.5}$$

Since $T_o \subset T_1 \subset \ldots \subset T_r = H \times H$ to prove the theorem it is enough to show that (1.3.4) holds on T_r.

The proof will be performed by induction with respect to q.

On the set T_o condition of compatibility is satisfied trivially, hence (1.3.4) holds by assumption.

Supposing that (1.3.4) holds on T_{q-1} we shall show that it is satisfied on T_q.

Let $(g,h) \in T_q$. Note that for any $g',h' \in \overline{[g,h]}$ we have $(g',h') \in T_q$. The interval $\overline{[g,h]}$ can be split into a finite number of subintervals $\overline{[g',h']}$ such that $|g'-h'| \leqslant \eta$. Obviously it is enough to show that (1.3.4) holds on each subinterval $\overline{[g',h']}$.

First let us assume that there exists a point $\delta_s \in \overline{[g',h']}$ such that

$$\#I(\delta_s) = q \tag{1.3.6}$$

If follows from the definition (1.3.5) that for every point s for which $\#I(\delta_s) = q$ we have $I(\delta_s) = J$, where J is the same for the whole interval $\overline{[g',h']}$.

Let us denote

$$\mu = \inf\{s \in [0,1] \mid I(\delta_s) = q\},$$
$$\nu = \sup\{s \in [0,1] \mid I(\delta_s) = q\}. \tag{1.3.7}$$

The property (1.3.1) implies that

$$I(\delta_\mu) = I(\delta_\nu) = J.$$

Moreover by the definition (1.3.5)

$$I(\delta_s) \subset J \qquad \forall s \in [0,1].$$

Hence (δ_μ, δ_ν) is a compatible pair. Therefore by the assumption

$$\|k(\delta_\nu) - k(\delta_\mu)\| \leqslant \gamma \|\delta_\nu - \delta_\mu\| = \gamma(\nu-\mu)\|h'-g'\| \tag{2.3.8}$$

Now let us consider the segment $(0,\mu)$(similarly $(\nu,1)$).

For any arbitrary $s \in (0,\mu)$ we have

$$I(\delta_s) < q. \tag{1.3.9}$$

On the other hand (1.3.1) implies that for every $s \in (0,\mu)$ there exists an open segment $\Delta_s \subset (0,\mu)$ containing s, such that for every $\sigma \in \Delta_s$ we have

$$I(\delta_\sigma) \subset I(\delta_s) . \tag{1.3.10}$$

Hence by the induction hypothesis and by (1.3.9) and (1.3.10) we get

$$|k(\delta_{\sigma_2})-k(\delta_{\sigma_1})| \leq \gamma|\delta_{\sigma_2}-\delta_{\sigma_1}| = \gamma|\sigma_2-\sigma_1||h'-g'| \quad \forall \sigma_1,\sigma_2 \in \Delta_s. \tag{1.3.11}$$

Let μ_1, μ_2 be arbitrary constants such that

$$0 < \mu_1 < \mu_2 < \mu.$$

The interval $[\mu_1,\mu_2]$ can be covered by the segments Δ_s constructed as above. From this covering we can extract a finite one. It allows to split the whole interval $[\mu_1,\mu_2]$ into a finite number of subintervals

$$\mu_1=\mu_1^1 < \mu_1^2 < \ldots < \mu_1^r=\mu_2$$

such that for each subinterval $[\mu_1^j,\mu_1^{j+1}]$ (1.3.11) holds.
Hence

$$|k(\delta_{\mu_2})-k(\delta_{\mu_1})| \leq \gamma|\delta_{\mu_2}-\delta_{\mu_1}| = \gamma|\mu_2-\mu_1||h'-g'| . \tag{1.3.12}$$

Letting $\mu_1 \to 0$, $\mu_2 \to \mu$ and taking advantage of continuity of k we get

$$|k(\delta_\mu)-k(\delta_0)| \leq \gamma|\delta_\mu - \delta_0| = \gamma\cdot\mu|h'-g'|. \tag{1.3.13}$$

In exactly the same way we get

$$|k(\delta_1)-k(\delta_\nu)| \leq \gamma\ |\delta_1-\delta_\nu| = \gamma(1-\nu)|h'-g'| \tag{1.3.14}$$

Combining (1.3.8), (1.3.13) and (1.3.14) yields the desired result

$$|k(h')-k(g')| \leq |k(\delta_1)-k(\delta_\nu)| + |k(\delta_\nu)-k(\delta_\mu)| + |k(\delta_\mu)-k(\delta_0)| \leq \gamma|h'-g'|. \tag{1.3.15}$$

Finally note that in case where there does not exist such a point $\delta_s \in \overline{[g',h']}$ for which (1.3.6) holds, we can repeat the proof of (1.3.13) for the whole interval $\overline{[g',h']}$ getting (1.3.13) with $\mu=1$.

This completes the proof. □

1.4. Lipschitz Continuity of Solutions to Convex Programming Problems

Our purpose is to prove that the solutions $u(.)$ of (P_h) and the associated multipliers $\lambda(.)$ are Lipschitz continuous functions of h on any compact and convex subset $\mathcal{K} \subset H$.

To this end we apply Theorem 1.2 putting

$$X=R^m,\ Y=R^{n+r},\ k^T(h)=[u^T(h),\lambda^T(h)],\quad I(h)=I_h.$$

Since by Theorem 1.1 $k(.)$ is a continuous function and by (1.2.7) condition (1.3.1) is satisfied, then to prove Lipschitz continuity of $k(.)$ on $\mathcal{H}$ it is enough to show that $k(.)$ is Lipschitz continuous with the same modulus on any compatible pair of points g,h close enough each to the other.

Let $(g,h) \in \mathcal{H} \times \mathcal{H}$ be a compatible pair such that

$$|g-h| \leqslant \eta, \tag{1.4.1}$$

where $\eta > 0$ will be defined later on.

By definition of a compatible pair we have

$$I_g = I_h = J, \tag{1.4.2}$$

Hence from (1.1.6) and (1.1.14) we find that at h and g the following equation holds

$$F(w(h),h) \overset{\text{def}}{=} \begin{bmatrix} D_u f(u(h),h)+D_u\phi_J^T(u(h),h)\lambda_J(h) \\ \phi_J(u(h),h) \end{bmatrix} = 0 , \tag{1.4.3}$$

where

$$w^T(h) \overset{\text{def}}{=} [u^T(h),\lambda_J^T(h)].$$

Let us define

$$h(s) = (1-s)h + sg ,$$

and introduce an auxiliary equation

$$\overline{F}(\overline{w},s) \overset{\text{def}}{=} F(\overline{w},h(s)) \qquad 0 \leqslant s \leqslant 1 . \tag{1.4.4}$$

Using implicit function theorem we shall show that for $s \in [0,1]$ the solution $\overline{w}(s)$ of (1.4.4) exists, is unique and moreover it is a differentiable function of s on (0,1), provided that $\eta > 0$ in (1.4.1) is small enough.

At $s=0$ and $s=1$ we have

$$\bar{w}(0)=w(h), \quad \bar{w}(1)=w(g). \tag{1.4.5}$$

which will allow to derive the needed properties of $w(.)$ from properties of $\bar{w}(.)$.
To apply implicit function theorem we must find $D_w\bar{F}(\bar{w},s)$ and $D_s\bar{F}(\bar{w},s)$.
From (1.4.3) and (1.4.4) we get

$$D_w\bar{F}(\bar{w},s)=\begin{bmatrix} D^2_{uu}f(\bar{u},h(s))+D^2_{uu}\phi^T_J(\bar{u},h(s))\bar{\lambda}_J, & D_u\phi^T_J(\bar{u},h(s)) \\ D_u\phi^T_J(\bar{u},h(s)) & , 0 \end{bmatrix} \tag{1.4.6}$$

and

$$D_s\bar{F}(\bar{w},s)=\begin{bmatrix} \langle D^2_{uh}f(\bar{u},h(s))+D^2_{uh}\phi^T_J(\bar{u},h(s))\bar{\lambda}_J, \ g-h\rangle \\ \langle D_h\phi_J(\bar{u},h(s)), \ g-h\rangle \end{bmatrix}. \tag{1.4.7}$$

Note that the matrix $D_w\bar{F}(\bar{w},s)$ has the following structure

$$A = \begin{bmatrix} B & , C^T \\ C & , 0 \end{bmatrix}. \tag{1.4.8}$$

We shall need the following simple result concerning the matrix A:

<u>Lemma 1.2</u>

If there exist positive constants ρ and γ such that

$$(x,Bx) \geq \rho|x| \qquad \forall x \tag{1.4.9}$$

and

$$|C^Ty| \geq \gamma|y|, \qquad \forall y \tag{1.4.10}$$

then the matrix A is non-singular and

$$|A^{-1}| \leq \max\left\{\frac{1}{\rho}, \frac{1}{2\gamma}\left[\sqrt{|B|^2+4\gamma}+|B|\right]\right\}. \tag{1.4.11}$$

<u>Proof</u>

Let μ denote any eigenvalue of A. We shall show that

$$|\mu| \geq \min\left\{\rho, \frac{2\gamma}{\sqrt{|B|^2+4\gamma}+|B|}\right\} \tag{1.4.12}$$

which implies (1.4.11).

Let $z^T=(x^T,y^T)$ be an eigenvector corresponding to μ, i.e.

$$Az = \mu Ez$$

or

$$(B-\mu E)x + C^T y = 0,$$

$$Cx - \mu Ey = 0 ,$$

where E denotes the unit matrix.

Note that if μ is an eigenvalue of B, then (1.4.9) implies (1.4.12). Hence it is enough to consider the case where $(B-\mu E)$ is non-singular. In this case we get

$$C(B - \mu E)^{-1}C^T y + \mu Ey = 0. \tag{1.4.13}$$

Note that if $0 < \mu < \rho$ then by (1.4.9) and (1.4.10) both matrices $C(B-\mu E)^{-1}C^T$ and μE are positive definite, hence (1.4.13) can not be satisfied and we get

$$\mu \geqslant \rho . \tag{1.4.14}$$

It remains to consider the case $\mu \leqslant 0$.
From (1.4.9) and (1.4.10) we get

$$y^T C(B - \mu E)^{-1}C^T y \geqslant \frac{\gamma}{|B|-\mu} .$$

Therefore a necessary condition for (1.4.13) to hold is that

$$-\mu \geqslant \frac{\gamma}{|B|-\mu} .$$

This implies

$$-\mu \geqslant \frac{1}{2} \left[\sqrt{|B|^2+4\gamma}-|B|\right] = \frac{2\gamma}{\sqrt{|B|^2+4\gamma}+|B|} . \tag{1.4.15}$$

(1.4.14) and (1.4.15) yield (1.4.12). □

Assumption (1.1.7) and Corollary 1.1 imply that

$$|C^T y| \overset{\text{def}}{=} |D_u \phi_J^T(\bar{u},h(s))y| \geqslant \frac{\beta}{2} |y| , \tag{1.4.16}$$

provided that

$$|u(h)-\bar{u}| \leqslant \xi , \quad |h-h(s)| \leqslant \zeta , \tag{1.4.17}$$

where ξ and ζ are constants independent of $h \in \mathcal{H}$.

Denote

$$B = B_1 + B_2 , \tag{1.4.18}$$

where

$$B_1 \overset{\mathrm{def}}{=} D^2_{uu} f(\bar{u},h(s)) + D^2_{uu}\phi^T_J(u,h(s))\, \lambda_J(h), \qquad (1.4.18a)$$

$$B_2 \overset{\mathrm{def}}{=} D^2_{uu}\phi^T_J(u,h(s))(\bar{\lambda}_J - \lambda_J(h)). \qquad (1.4.18b)$$

By (A1), (A3) and (1.1.16) we get

$$< x, B_1 x > \geq \alpha |x|^2 \qquad \forall x \in R^n .$$

On the other hand by Corollary 1.1 there exists a constant $\theta > 0$, independent of $h \in \mathcal{H}$ such that

$$|B_2| \leq \frac{\alpha}{2},$$

provided that (1.4.17) holds and

$$|\bar{\lambda}_J - \lambda_J(h)| \leq \theta. \qquad (1.4.19)$$

In that case we get

$$< x, Bx > \geq \frac{\alpha}{2} |x|^2 . \qquad (1.4.20)$$

On the other hand if (1.4.17) and (1.4.19) hold then

$$|B| \leq |B_1| + |B_2| \leq M, \qquad (1.4.21)$$

where M is a constant independent of $h \in \mathcal{H}$.
By (1.4.6), (1.4.16), (1.4.20) and (1.4.21) as well as by Lemma 1.2 we get

$$|D_W \bar{F}^{-1}(\bar{w},s)| \leq c_1 \qquad (1.4.21)$$

where c_1 is independent of $h \in \mathcal{H}$, provided that (1.4.17) and (1.4.19) hold.

Under the same assumptions from (1.4.7) we obtain

$$|D_S \bar{F}(\bar{w},s)| \leq c_2 |g-h| , \qquad (1.4.22)$$

where c_2 is independent of $h \in \mathcal{H}$.

Applying implicit function theorem (see e.g. [28]) to (1.4.4) we find

$$D_S \bar{w}(s) = -[D_W \bar{F}(w(s),s)]^{-1} [D_S \bar{F}(w(s),s)]. \qquad (1.4.23)$$

Hence by (1.4.21), (1.4.22) we get

$$|D_s\bar{w}(s)| \leq c|g-h| , \qquad (1.4.24)$$

where c is independent of $h \in \mathcal{H}$.

From (1.4.5) and (1.4.23) we obtain

$$|w(g)-w(h)|=|\bar{w}(1)-\bar{w}(0)|=|\int_0^1 D_s\bar{w}(s)ds| \leq c|g-h|. \qquad (1.4.25)$$

Using definition of $w(.)$ and taking into account that by (1.4.2)

$$\lambda^i(g) = \lambda^i(h) \quad \text{for} \quad i \notin J$$

we finally obtain

$$|u(h)-u(g)|, \ |\lambda(h)-\lambda(g)| \leq c|g-h| \qquad (1.4.26)$$

where c does not depend on $h \in \mathcal{H}$.

Inequality (1.4.26) holds provided that conditions (1.4.17) and (1.4.19) are satisfied. By (1.4.26) these conditions are satisfied if we put in (1.4.1)

$$\eta = \min \{\zeta, \frac{\xi}{c} , \frac{\theta}{c} \} \qquad (1.4.27)$$

Note that η does not depend on $h \in \mathcal{H}$.
Hence by Theorem 1.2 we get:

Theorem 1.3

If assumptions (A1) through (A6) hold, then solutions $u(h)$ and associated Lagrange multipliers $\lambda(h)$ are Lipschitz continuous functions of h on any compact and convex set $\mathcal{H} \subset H$.

Remark 1.1

For the sake of simplicity it was assumed that in Problem (P_h) only inequality type constraints are present. However all the above results, including Theorem 1.3, remains valued if Φ_h containes affine equality type constraints, provided that (A6) holds for all binding constraints, including those of equality type.

2. CONVEX OPTIMAL CONTROL PROBLEM SUBJECT TO CONTROL CONSTRAINTS

2.1. Problem Statement

This chapter is devoted to an abstract problem of optimization, which is a model for a class of convex optimal control problems subject to pointwise control constraints.

Like in Chapter 1 $H \subset R^m$ denotes an open and convex set.

We introduce a Hilbert space U (space of control) and a pair of other Hilbert spaces

$$Y \subset Z ,$$

where in applications Z will be the space of output and Y - the space of state.

The inner products and the norms in Y and Z will be denoted by $(.,.)_Y$, $(.,.)_Z$ and $||\cdot||_Y$, $||\cdot||_Z$ respectively.

For each $h \in H$ there are given

- a convex functional

$$F(.,.,h) : U \times Z \to R^1 ,$$

- a linear, continuous mapping from U into Y (given by the state equation)

$$S(h) \in \mathcal{L}(U ; Y) , \qquad (2.1.1)$$

- a closed and convex subset $U_h^{ad} \subset U$ (set of admissible control).

We consider a family $\{0_h\}$ of the following convex problems of optimization

$$(0_h) \quad \left| \begin{array}{ll} \text{find a pair } (u_h, z_h) \in U \times Z, \text{ such that} & \\ F(u_h, z_h, h) = \min\limits_{u \in U_h^{ad}} F(u,z,h) & (2.1.2) \\ \text{subject to} & \\ z = S(h)u. & (2.1.3) \end{array} \right.$$

In the sequel we shall put

$$U = L^2(\Xi;R^n) ,$$

where Ξ is a given bounded subset of R^q, and $L^2(\Xi;R^n)$ is the space of R^n - valued functions square integrable on Ξ.

$$U_h^{ad} = \{u \in L^2(\Xi;R^n) | u(\xi) \in \Phi_h \quad \text{a.a. } \xi \in \Xi\}, \tag{2.1.4}$$

where

$$\Phi_h = \{u \in R^n | \phi(u,h) \leqslant 0\}, \tag{2.1.5}$$

$$\phi(u,h) = [\phi^1(u,h), \phi^2(u,h), \ldots, \phi^r(u,h)]^T ,$$

$$F(u,z,h) = F^1(u,h) + F^2(z,h) \tag{2.1.6}$$

where

$$F^1(u,h) = \int_\Xi f^1(u(\xi),h)d\xi. \tag{2.1.7}$$

We assume that the following conditions hold:

(B1) the embeding $Y \subset Z$ is continuous and compact,

(B2) for each $h \in H$ the functionals $F^1(.,h)$ and $F^2(.,h)$ are convex and twice continuously differentiable on U and Z respectively. Moreover there exists a constant $\alpha > 0$ independent of h, such that

$$< v, D_{uu}^2 f^1(u,h)v > \geqslant \alpha|v|^2 \qquad \forall u,v \in R^n, \quad \forall h \in H, \tag{2.1.8}$$

(B3) the functions $F^1(.,.)$, $D_u F^1(.,.)$ are continuously differentiable on $U \times H$, while $F^2(.,.)$, $D_z F^2(.,.)$ are continuously differentiable on $Z \times H$ and on $Y \times H$,

(B4) for each $h \in H$

$$S(h) \in \mathcal{L}(L^2(\Xi;R^n);Y)$$

is a linear, continuous operator from $U=L^2(\Xi;R^n)$ into Y, while

$$S^*(h) \in \mathcal{L}(Z;L^2(\Xi;R^n)) \cap \mathcal{L}(Y;L^4(\Xi;R^n)),$$

(B5) $S(.)$ and $S^*(.)$ are continuously differentiable fuctions from H into $\mathcal{L}(L^2(\Xi;R^n);Y)$ and $\mathcal{L}(Z;L^2(\Xi;R^n)) \cap \mathcal{L}(Y;L^4(\Xi;R^n))$ respectively,

(B6) for each $h \in H$ $\phi^i(.,h)$ $(i=1,2,\ldots,r)$ are twice continuously differentiable and convex functions on R^n,

(B7) the functions $\phi^i(.,.)$ and $D_u\phi^i(.,.)$ $(i=1,2,\ldots,r)$ are continuously differentiable on $R^n \times H$,

(B8) for all $h \in H$ the admissible sets Φ_h are non-empty

$$\Phi_h \neq \emptyset \qquad \forall h \in H$$

and uniformly bounded, i.e. there exists a constant c independent of h, such that

$$|u| \leq c \qquad \forall u \in \Phi_h, \quad \forall h \in H. \tag{2.1.9}$$

It is well known [35] that under assumptions (B2), (B4), (B6) and (B8) Problem (0_h) has a unique solution (u_h, z_h).

In the sequel we denote by $(.,.)$ the inner product in L^2 space while $||\cdot||$ and $||\cdot||_4$ - denote the norms in L^2 and L^4 respectively.

$||\cdot||_{\mathcal{L}(U;Z)}$ - denotes the norm in the operator space $\mathcal{L}(U;Z)$.

Moreover let us denote

$$I = \{1,2,\ldots,r\}$$

and introduce the sets

$$J_h(u) = \{i \in I \mid \phi^i(u,h) = 0\}, \tag{2.1.10a}$$

$$I_h(\xi) = \{i \in I \mid \phi^i(u_h(\xi),h) = 0\}, \tag{2.1.10b}$$

of all indices of the control functions binding at some $u \in R^n$ and at $u_h(\xi)$ respectively.

In addition to (B1) through (B8) we assume that for all $h \in H$ and for all $u \in U_h^{ad}$ the constraint regularity condition analogous to (A6) holds:

(B9) there exists a constant $\beta > 0$ such that

$$|[D_u \phi^T_{J_h(u)}(u,h)]v| \geq \beta |v| \tag{2.1.11}$$

for each $h \in H$, each $u \in U_h^{ad}$ and each v of appropriate dimension, where $D_u \phi^T_{J_h(u)}(u,h)$ denotes the matrix whose columns are the gradients of all constraint functions $\phi^i(.,h)$ binding at u.

Remark 2.1

For the sake of simplicity in the formulation of Problem (0_h) it was assumed that the functional F has the form (2.1.6).

It can contain also terms depending both on u and x as it will be done in Chapter 3. Also the linear mapping $S(h)$ can be substituted by an affine one. On the other hand the pointwise character of control constraints and condition (B9) are essential in our considerations.

Let us introduce Lagrangian

$$\bar{L}(.,.;.;.) : L^2(\Xi;R^n) \times Z \times Z \times H \to R^1$$

$$\bar{L}(u,z;p;h) \overset{def}{=} F(u,z,h) + (p,z-S(h)u)_Z. \tag{2.1.12}$$

It is well known [35] that there exists a unique Lagrange multiplier p_h such that (u_h,z_h) is characterized by the degenerated saddle point of Lagrangian $\bar{L}$. Namely

$$\bar{L}(u_h,z_h;p;h) = \bar{L}(u_h,z_h;p_h;h) \leqslant \bar{L}(u,z;p_h;h) \tag{2.1.13}$$

$$\forall u \in U_h^{ad} \quad , \quad \forall z,p \in Z.$$

Condition (2.1.13) is equivalent to the following ones

$$p_h = -D_z F^2(z_h,h) \tag{2.1.14}$$

$$(D_u F^1(u_h,h) - S^*(h)p_h, u-u_h) \geqslant 0 \qquad \forall u \in U_h^{ad} . \tag{2.1.15}$$

Note that by (2.1.3), (2.1.14) as well as by (B5) and (B8) it follows that z_h, p_h are bounded in Y norm uniformly on any compact subset $\mathcal{H} \subset H$:

$$||z_h||_Y, ||p_h||_Y \leqslant c \qquad \forall h \in \mathcal{H} . \tag{2.1.16}$$

Taking advantage of the form (2.1.7) of $F^1(u,h)$ and of the fact that the control constraints are of the pointwise type (2.1.4) we find that (2.1.15) is equivalent to

$$\langle D_u f^1(u_h(\xi),h)-(S^*(h)p_h)(\xi), v-u_h(\xi)\rangle \geqslant 0 \quad \text{for all} \quad v \in \phi_h,$$

$$\text{for a. a.} \quad \xi \in \Xi. \tag{2.1.17}$$

Note that $u_h(\xi)$ satisfies (2.1.17) if and only if it is the solution of the following convex programming problem

$$(CP_h) \quad \left| \begin{array}{l} f^1(u_h(\xi),h) - \langle (S^*(h)p_h)(\xi), u_h(\xi) \rangle = \\ \\ = \min\limits_{v \in \Phi_h} \{f^1(v,h) - \langle (S^*(h)p_h)(\xi), v \rangle\}. \end{array} \right. \tag{2.1.18}$$

By (B2), (B6), (B8) and (B9) conditions (A.1), (A3), (A5) and (A6) of Section 1.1 are satisfied and like in the case of Problem (P_h) there exists a uniquely defined Lagrange multiplier $\lambda_h(\xi) \in R^r$ such that $u_h(\xi)$ is characterized by the Kuhn-Tucker conditions

$$D_u f^1(u_h(\xi),h)-(S^*(h)p_h)(\xi)+D_u\phi^T((u_h(\xi),h)\lambda_h(\xi) = 0 , \tag{2.1.19}$$

$$\lambda_h^i(\xi)\phi^i(u_h(\xi),h) = 0 \;, \tag{2.1.20a}$$

$$\lambda_h^i(\xi) \geq 0 \qquad i \in I \;. \tag{2.1.20b}$$

The function $\lambda_h(.)$ is well defined by (2.1.19), (2.1.20) almost everywhere on Ξ.

Lemma 2.1

The function λ_h belongs to $L^4(\Xi;R^r)$, and for every compact subset $\mathcal{H} \subset H$ there exists a constant c such that

$$||\lambda_h||_4 \leq c \qquad \forall h \in \mathcal{H} \tag{2.1.21}$$

Proof

First we are going to show that $\lambda_h(.)$ is measurable. Let $K \subset I$ be any arbitrary subset of indices. Denote

$$\Xi_{h,K} = \{\xi \in \Xi \,|\, I_h(\xi) = K\}.$$

It is clear that the sets $\Xi_{h,K}$ are measurable,

$$\Xi_{h,K_1} \cap \Xi_{h,K_2} = \emptyset \quad \text{for} \quad K_1 \neq K_2$$

and

$$\text{meas} \bigcup_{K \subset I} \Xi_{h,K} = \text{meas}\; \Xi,$$

hence it is enough to show that λ_h is measurable on any $\Xi_{h,K}$.

Let us denote by $\lambda_{h,I_h(\xi)}(\xi)$ the subvector of $\lambda_h(\xi)$ containing all components $\lambda_h^i(\xi)$ such that $i \in I_h(\xi)$. Hence for all $\xi \in \Xi_{h,K}$, with fixed K, the vectors $\lambda_{h,I_h(\xi)}(\xi)$ contain the same components.

Using (2.1.20a) we can rewrite (2.1.19) in the form

$$D_u f^1(u_h(\xi),h)-(S_h^* p_h)(\xi)+D_u \phi^T_{I_h(\xi)}(u_h(\xi),h)\lambda_{h,I_h(\xi)}(\xi)=0. \tag{2.1.22}$$

Note that by (2.1.10) there exists a left-inverse $\mathcal{K}(\xi)$ of the matrix $D_u\phi^T_{I_h(\xi)}(u_h(\xi),h)$. Hence

$$\lambda_{h,I_h(\xi)}(\xi)=\mathcal{K}(\xi)\left[-D_u f^1(u_h(\xi),h)+(S^*(h)p_h)(\xi)\right]. \tag{2.1.23}$$

It is easy to see that $\varkappa(\xi)$ can be chosen in such a way that $\varkappa(.)$ is a function measurable on $\Xi_{h,K}$, therefore by (2.1.23) $\lambda_{h,I_h(\xi)}$ is also measurable on $\Xi_{h,K}$.

Since by (2.1.20a) $\lambda^i(\xi)=0$, for $i\notin K$ and for all $\xi\in\Xi_{h,K}$, then we find that λ_h is measurable on $\Xi_{h,K}$, i.e. it is also measurable on Ξ.

To prove (2.1.21) let us note that by (2.1.11) and (2.1.20a) we get from (2.1.22)

$$|\lambda_h(\xi)|=|\lambda_{h,I_h(\xi)}(\xi)| \leq \frac{1}{\beta}\,[|D_u f^1(u_h(\xi),h)|+|(S^*(h)p_h)(\xi)|]$$

or

$$|\lambda_h(\xi)|^4 \leq \frac{8}{\beta^4}\,[|D_u f^1(u_h(\xi),h)|^4+|(S^*(h)p_h)(\xi)|^4]. \qquad (2.1.24)$$

Taking into account (B3), (B4), (B8) and (2.1.16) and integrating (2.1.24) over Ξ we obtain (2.1.21).

□

Now we can augment Lagrangian (2.1.12) adding the term corresponding to the constraints (2.1.5). In this way we get a new Lagrangian

$$L(.,.;.,.;.) : L^2(\Xi;R^n) \times Z \times Z \times L^2(\Xi;R^r) \times H \to R^1$$

$$L(u,z;p,\lambda;h)=\bar{L}(u,z,p,h)+(\lambda,\phi(u,h))=F(u,z,h)+(p,z-S(h)u)_Z+(\lambda,\phi(u,h)). \qquad (2.1.25)$$

From (2.1.13), (2.1.19) and (2.1.20) it follows that Lagrangian (2.1.25) assumes its saddle point at $(u_h,z_h;p_h,\lambda_h)$ i.e.

$$L(u_h,z_h;p\ ,\lambda;h) \leq L(u_h,z_h;p_h,\lambda_h;h) \leq L(u,z;p_h,\lambda_h;h) \qquad (2.1.26)$$

$$\forall u\in L^2(\Xi;R^n),\ \forall z,p\in Z,\ \forall\lambda\in L^2(\Xi;R^r),\ \lambda(\xi)\geq 0 \text{ for a.a. } \xi\in\Xi.$$

From (2.1.26) we get the following differential conditions

$$D_u L(u_h,z_h;p_h,\lambda_h;h) = 0, \qquad (2.1.27a)$$

$$D_z L(u_h,z_h;p_h,\lambda_h;h) = 0. \qquad (2.1.27b)$$

2.2. Lipschitz Continuity with Respect to Parameter

In this section we are going to show Lipschitz continuity with respect to the parameter of primal and dual optimal variables for (O_h). Namely we prove the following theorem:

Theorem 2.1

If conditions (B.1) through (B.9) hold then for any compact and convex set $\mathcal{H} \subset H$ there exists a constant c such that

$$||u_2-u_1||, ||z_2-z_1||_Y, ||p_2-p_1||_Y, ||\lambda_2-\lambda_1|| \leq c|h_2-h_1|, \tag{2.2.1}$$

where c does not depend on $h \in \mathcal{H}$, and on the left-hand side the subscripts 1 and 2 are used instead of h_1 and h_2.

Proof

Let us choose any arbitrary $h_1, h_2 \in \mathcal{H}$.
Expanding $L(\cdot,\cdot,p_2,\lambda_2,h_1)$ into Taylor's series at (u_2,z_2) and taking advantage of (2.1.8) we get

$$L(u_1,z_1;p_2,\lambda_2;h_1) \geq L(u_2,z_2;p_2,\lambda_2;h_1)+(D_uL(u_2,z_2,\lambda_2;h_1), u_1-u_2)+$$
$$+ (D_zL(u_2,z_2;p_2,\lambda_2;h_1),z_1-z_2)_Z+ \frac{\alpha}{2}||u_2-u_1||^2. \tag{2.2.2}$$

From (2.1.26) it follows that

$$L(u_1,z_1;p_2,\lambda_2;h_1) \leq L(u_2,z_2;p_1,\lambda_1;h_1). \tag{2.2.3}$$

Substituting (2.2.3) into (2.2.2) yields

$$||u_2-u_1||^2 \leq c\{[L(u_2,z_2;p_1,\lambda_1;h_1)-L(u_2,z_2;p_2,\lambda_2;h_1)] +$$
$$+(D_uL(u_2,z_2;p_2,\lambda_2;h_1),u_2-u_1)+(D_zL(u_2,z_2;p_2,\lambda_2;h_1),z_2-z_1)_Z\}. \tag{2.2.4}$$

We are going to estimate all three terms on the right-hand side of (2.2.4).

Using definition (2.1.25) we get

$$L(u_2,z_2;p_1,\lambda_1;h_1)-L(u_2,z_2;p_2,\lambda_2;h_1)=(p_1-p_2,z_2-S(h_1)u_2)_Z+(\lambda_1-\lambda_2,\phi(u_2,h_1)).$$

On the other hand

$$z_2 - S(h_2)u_2 = 0,$$

and by (2.1.20)

$$(\lambda_1-\lambda_2,\phi(u_2,h_2)) \leq 0.$$

Hence taking into account (B.5) and (B.7) we get

$$L(u_2,z_2;p_1,\lambda_1;h_1)-L(u_2,z_2;p_2,\lambda_2;h_1) \leq$$

$$\leq (p_1-p_2,(S(h_2)-S(h_1))u_2)_Z+(\lambda_1-\lambda_2,\phi(u_2,h_1)-\phi(u_2,h_2)) \leq$$

$$\leq ||p_1-p_2||_Z||S(h_1)-S(h_2)||_{\mathcal{L}(U,Z)}||u_2||+||\lambda_1-\lambda_2|| \, ||\phi(u_2,h_1)-\phi(u_2,h_2)|| \leq$$

$$\leq c \, |h_2-h_1| \, (||p_2-p_1||_Z+||\lambda_2-\lambda_1||). \tag{2.2.5}$$

Let us estimate two remaining terms on the right-hand side of (2.2.4). Taking advantage of (2.1.27a) as well as of (B.1), (B.3), (B.5), (B.7), (2.1.16), (2.1.21) and (2.1.27a) we obtain

$$(D_uL(u_2,z_2;p_2,\lambda_2;h_1),u_2-u_1) =$$

$$=(D_uL(u_2,z_2;p_2,\lambda_2;h_1)-D_uL(u_2,z_2;p_2,\lambda_2;h_2),u_2-u_1) =$$

$$=([D_uF^1(u_2,h_1)-D_uF^1(u_2,h_2)]+(S^*(h_1)-S^*(h_2))p_2 +$$

$$+(D_u\phi^*(u_2,h_1)-D_u\phi^*(u_2,h_2))\lambda_2,u_2-u_1) \leq$$

$$\leq [\, ||D_uF^1(u_2,h_1)-D_uF^1(u_2,h_2)||+||S^*(h_1)-S^*(h_2)||_{\mathcal{L}(U;Z)}||p_2||_Z+$$

$$+ ||D_u\phi(u_2,h_1)-D_u\phi(u_2,h_2)|| \, ||\lambda_2||\,] \, ||u_2-u_1|| \leq$$

$$\leq c|h_2-h_1| \, ||u_2-u_1||. \tag{2.2.6}$$

Similarly

$$(D_zL(u_2,z_2;p_2,\lambda_2;h_1),z_2-z_1)_Z=$$

$$=(D_zL(u_2,z_2;p_2,\lambda_2;h_1)-D_zL(u_2,z_2;p_2,\lambda_2;h_2),z_2-z_1)_Z \leq$$

$$\leq ||D_zF^2(z_2,h_1)-D_zF^2(z_2,h_2)||_Z||z_2-z_1||_Z \leq c|h_2-h_1| \, ||z_2-z_1||_Z. \tag{2.2.7}$$

Substituting (2.2.5) through (2.2.7) into (2.2.4) yields

$$||u_2-u_1||^2 \leq c|h_2-h_1|(||u_2-u_1||+||z_2-z_1||_Z+||p_2-p_1||_Z+||\lambda_2-\lambda_1||), \tag{2.2.8}$$

where c does not depend on $h_1,h_2 \in \mathcal{H}$.

Now we shall estimate $||z_2-z_1||_Z$, $||p_2-p_1||_Z$ and $||\lambda_2-\lambda_1||$ in terms of $||u_2-u_1||$.

Using (B1), (B5) and (2.1.9) we get from (2.1.3)

$$||z_2-z_1||_Z \leq c||z_2-z_1||_Y = c||S(h_2)u_2-S(h_1)u_1||_Y \leq$$

$$\leq c\,[\,||S(h_2)||_{\mathcal{L}(U;Y)}||u_2-u_1||+||S(h_2)-S(h_1)||_{\mathcal{L}(U;Y)}||u_1||\,]\leq$$

$$\leq c\,[\,||u_2-u_1||+|h_2-h_1|\,]. \tag{2.2.9}$$

Similarly by (B1), (B3), (2.1.14) and (2.2.9)

$$||p_2-p_1||_Z \leq c||p_2-p_1||_Y = c||D_zF^2(z_1,h_1)-D_zF^2(z_2,h_2)||_Y \leq c[\,||u_2-u_1||+|h_2-h_1|\,]. \tag{2.2.10}$$

Finally to estimate $||\lambda_2-\lambda_1||$ we need the following:

Lemma 2.2

The following estimates take place

$$|u_2(\xi)-u_1(\xi)|,\ |\lambda_2(\xi)-\lambda_1(\xi)| \leq c''(\xi)|h_2-h_1| +$$

$$+ c'(\xi)|(S^*(h_2)p_2)(\xi)-(S^*(h_1)p_1)(\xi)| \quad \text{for almost all } \xi\in\Xi, \tag{2.2.11}$$

where $c'\in L^2(\Xi;R^1)$, $c''\in L^4(\Xi;R^1)$ and there exists a constant c independent of $h_1,h_2\in\mathcal{H}$ such that

$$||c'||_4 \leq c\ ,\ ||c''|| \leq c. \tag{2.2.12}$$

Proof

Let us denote

$$k(s) \overset{\text{def}}{=} (1-s)(S^*(h_1)p_1)(\xi)+s(S^*(h_2)p_2)(\xi) \tag{2.2.13a}$$

$$h(s) \overset{\text{def}}{=} (1-s)h_1+sh_2 \qquad s\in[0,1] \tag{2.2.13b}$$

and consider an auxiliary convex programming problem ($\overline{CP}_s$) depending on a real-valued parameter $s\in[0,1]$

$$(\overline{CP}_s)\quad \left|\begin{array}{l} \text{find } v(s)\in R^n \text{ such that} \\ \\ f^1(v(s),h(s))-\langle k(s),v(s)\rangle = \min\limits_{v\in\Phi_{h(s)}} \{f^1(v,h(s))-\langle k(s),v\rangle\} \end{array}\right. \tag{2.2.14}$$

For each $s \in [0,1]$ Problem $(\overline{CP}_s)$ has a unique solution and at $s=0$ and $s=1$ $(\overline{CP}_s)$ coincides with $(\overline{CP}_h)$ at h_1 and h_2 respectively.

Hence

$$v(0) = u_1(\xi) \quad \text{and} \quad v(1) = u_2(\xi). \tag{2.2.15a}$$

Similarly

$$\nu(0) = \lambda_1(\xi) \quad \text{and} \quad \nu(1) = \lambda_2(\xi), \tag{2.2.15b}$$

where $\nu(s)$ is the Lagrange multiplier associated with $(\overline{CP}_s)$.

Taking into account (2.1.8) and (2.2.13a) and using the same argument as in the proof of Lemma 1.1 we get

$$|\nu(s)| \leq \frac{1}{\beta}\left[|f^1(v(s),h(s)|+|k(s)|\right] \leq$$

$$\leq \frac{1}{\beta}\left[|f^1(v(s),h(s)|+\max_{i=1,2}\{|(S^*(h_i)p_i)(\xi)|\}\right]. \tag{2.2.17}$$

It is easy to see that for $(\overline{CP}_s)$ all assumptions of Theorem 1.3 hold, hence by that theorem $v(.)$ and $\nu(.)$ are Lipschitz continuous functions on $[0,1]$. To prove our lemma we must find the corresponding Lipschitz modulus. To this end we use the same argument as in the proof of Theorem 1.3.

Namely we shall estimate the Lipschitz modulus uniformly for all compatible pairs (s_1,s_2) such that $|s_2-s_1|$ is sufficiently small.

Let (s_1,s_2) be a compatible pair, i.e.

$$I_{h(s_1)} = I_{h(s_2)} = K$$

and $I_{h(s)} \subset K$ for all $s \in (s_1,s_2)$.

Analogically to (1.4.3), (1.4.4) let us denote

$$F(\overline{w},k,h) \overset{\text{def}}{=} \begin{bmatrix} D_u f^1(u,h)-k-D_u\phi_K^T(u,h)\nu_K \\ \phi_K(u,h) \end{bmatrix}, \tag{2.2.18}$$

$$\overline{F}(\overline{w},s) \overset{\text{def}}{=} F(\overline{w},k(s),h(s)) \quad \text{for} \quad s \in [s_1,s_2], \tag{2.2.19}$$

where

$$\overline{w}^T \overset{\text{def}}{=} [\overline{v}^T,\overline{\nu}_K^T].$$

Using the implicit function theorem, in the same way as in the proof of Theorem 1.3, we obtain

$$|\bar{w}(s_2)-\bar{w}(s_1)|=|\int_{s_1}^{s_2} D_s\bar{w}(s)ds| \leq \int_{s_1}^{s_2}|[D_w\bar{F}(\bar{w}(s),s)]^{-1}|\,|D_s\bar{F}(\bar{w}(s),s)|\, ds \leq$$

$$\leq \max_{s\in[0,1]} |[D_w\bar{F}(\bar{w}(s),s)]^{-1}|\cdot \max_{s\in[0,1]} |D_s\bar{F}(\bar{w}(s),s)|\,|s_2-s_1|. \tag{2.2.20}$$

We are going to estimate ooth components on the right-hand side of (2.2.20). By (2.2.13), (2.2.18) and (2.2.19) we have

$$D_w\bar{F}(\bar{w},s) = \begin{bmatrix} D^2_{uu}f^1(\bar{v},h(s))+D^2_{uu}\phi_K^T(\bar{v},h(s))\bar{v}_K, & D_u\phi_K^T(\bar{v},h(s)) \\ D_u\phi_K(\bar{v},h(s)) & , 0 \end{bmatrix} \tag{2.2.21}$$

$$D_s\bar{F}(\bar{w},s) = \begin{bmatrix} < D^2_{uh}f^1(\bar{v},h(s))+D^2_{uh}\phi_K^T(\bar{v},h(s))\bar{v}_K, h_2-h_1> \\ < D_h\phi_K(\bar{v},h(s)),h_2-h_1> \end{bmatrix} + \begin{bmatrix} (S^*(h_1)p_1)(\xi)-(S^*(h_2)p_2)(\xi) \\ 0 \end{bmatrix}$$

(2.2.22)

Taking advantage of (B9) and of Lemma 1.2, and using the same argument as in the proof of Theorem 1.3 we find that if $|\bar{\nu}-\nu(s_1)|$ is sufficiently small, then

$$|[D_w\bar{F}(\bar{w},s)]^{-1}| \leq \max\{\frac{2}{\alpha}, \frac{1}{\beta}[(B^2+2\beta)^{1/2}+B]\}, \tag{2.2.23}$$

where

$$B = \max_{s\in[0,1]} \max_{\bar{v}\in\Phi_{h(s)}} |D^2_{uu}f^1(\bar{v},h(s))+D^2_{uu}\phi_K(\bar{v},h(s))\nu_K(s_1)|.$$

Using (2.1.9) and (2.2.17) we obtain

$$B \leq \max_{s\in[0,1]} \max_{\bar{v}\in\Phi_{h(s)}} \{|D^2_{uu}f(\bar{v},h(s))|+|D^2_{uu}\phi(\bar{v},h(s))|\,|\nu(s)|\} \leq$$

$$\leq c' + c'' \max_{i=1,2}\{|(S^*(h_i)p_i)(\xi)|\} \overset{\text{def}}{=} \rho(\xi). \tag{2.2.24}$$

Taking into account (B.4), (B.5) and (2.1.16), we find that the function ρ defined almost everywhere on Ξ by (2.2.24) is an element of $L^4(\Xi;R^1)$ and there is a constant $c>0$ independent of $h_1,h_2\in\mathcal{H}$ such that

$$||\rho||_4 \leq c.$$

Hence from (2.2.23) we obtain

$$|[D_w\bar{F}(\bar{w},s)]^{-1}| \leq \rho'(\xi), \tag{2.2.25}$$

where $\rho' \in L^4(\Xi;R^1)$ and

$$||\rho'||_4 \leq c \quad \text{for all} \quad h_1,h_2 \in \mathcal{H} . \tag{2.2.25a}$$

Similarly from (2.2.17) and (2.2.22) we get

$$|D_s\overline{F}(\overline{w},s)| \leq \max_{s\in[0,1]} \max_{\overline{v}\in\phi_h(s)} \{[|D^2_{uh}f^1(\overline{v},h(s))+D^2_{uh}\phi^T_K(\overline{v},h(s))\nu_K(s)|^2 +$$

$$+|D_h\phi_K(\overline{v},h(s))|^2]^{1/2}\}|h_2-h_1|+|(S^*(h_1)p_1)(\xi)-(S^*(h_2)p_2)(\xi)| \leq$$

$$\leq \rho''(\xi)|h_2-h_1|+|(S^*(h_1)p_1)(\xi)-(S^*(h_2)p_2)(\xi)|, \tag{2.2.26}$$

where

$$||\rho''||_4 \leq c \quad \text{for all} \quad h_1,h_2 \in \mathcal{H} . \tag{2.2.26a}$$

Substituting (2.2.25) and (2.2.26) into (2.2.20) we get

$$|\overline{w}(s_2)-\overline{w}(s_1)|\leq[\rho'(\xi)\rho''(\xi)|h_2-h_1|+\rho'(\xi)|(S^*(h_1)p_1)(\xi)-(S^*(h_2)p_2)(\xi)|]\cdot|s_2-s_1|$$

and in exactly the same way as in the proof of Theorem 1.3 we find that $\overline{w}(.)$ is Lipschitz continuous on $[0,1]$ whereas taking into account (2.2.15) we get

$$|u_2(\xi)-u_1(\xi)|,|\nu_2(\xi)-\nu_1(\xi)| \leq |\overline{w}(1)-\overline{w}(0)| \leq$$

$$\leq \rho'(\xi)\rho''(\xi)|h_2-h_1|+\rho'(\xi)|(S^*(h_1)p_1)(\xi)-(S^*(h_2)p_2)(\xi)|. \tag{2.2.27}$$

Hence (2.2.11) holds with

$$c'(\xi) = \rho'(\xi) \quad , \quad c''(\xi) = \rho'(\xi)\rho''(\xi),$$

while (2.2.25a) and (2.2.26a) imply (2.2.12). □

Using (B.4), (B.5), (2.1.16), (2.2.10) and (2.2.12) we obtain from (2.2.11)

$$||\lambda_2-\lambda_1|| \leq c\,[|h_2-h_1|+||S^*(h_2)p_2-S^*(h_1)p_1||] \leq c\,[||u_2-u_1||+|h_2-h_1|].$$

Substituting (2.2.9), (2.2.10) and (2.2.28) into (2.2.8) we get

$$||u_2-u_1|| \leq c|h_2-h_1|,$$

which together with (2.2.9), (2.2.10) and (2.2.28) complete the proof of the theorem. □

3. CONVEX OPTIMAL CONTROL PROBLEM SUBJECT TO STATE AND CONTROL CONSTRAINTS

3.1. Problem Statement

In this chapter we consider convex optimal control problems that depend on a parameter. The problems are subject to state and control constraints.

For each value of the parameter we investigate regularity of primal and dual optimal solutions and later on we show that these solutions are Lipschitz continuous functions of the parameter.

Our problems are formulated as follows.
As before $H \subset R^m$ denotes on open and convex set of vector parameters. Let us consider a family $\{OC_h\}$ of the following optimal control problems depending on h:

(OC_h)

find a pair $(u_h, x_h) \in L^2(0,T) \times C(0,T)$ such that

$$F(u_h, x_h, h) = \min\{F(u,x,h) \overset{\text{def}}{=} \int_0^T f(u(t),x(t),h)dt\} \tag{3.1.1}$$

subject to

$$\dot{x}(t) = A(h)x(t) + B(h)u(t), \tag{3.1.2}$$

$$x(0) = x^o \tag{3.1.2a}$$

and

$$u(t) \in \Phi_h \overset{\text{def}}{=} \{u \in R^n \mid \phi^i(u,h) \leq 0, i=1,2,\dots,r\} \text{ for a.a. } t \in [0,T], \tag{3.1.3}$$

$$x(t) \in \Theta_h \overset{\text{def}}{=} \{x \in R^\ell \mid \theta^j(x,h) \leq 0, j=1,2,\dots,s\} \text{ for all } t \in [0,T]. \tag{3.1.4}$$

Let us denote

$$\phi(u,h) = [\phi^1(u,h), \phi^2(u,h), \dots, \phi^r(u,h)]^T,$$

$$\theta(x,h) = [\theta^1(x,h), \theta^2(x,h), \dots, \theta^s(x,h)]^T,$$

and

$$I = \{1,2,\dots,r\}, \quad J = \{1,2,\dots,s\}.$$

We assume that the following conditions hold:

(C1) for each $h \in H$ the function $f(.,.,h)$ is convex and twice continuously differentiable in both variables. Moreover there exists a constant $\alpha > 0$ independent of h such that

$$[v^T,y^T]\begin{bmatrix} D^2_{uu}f(u,x,h), & D^2_{ux}f(u,x,h) \\ D^2_{xu}f(u,x,h), & D^2_{xx}f(u,x,h)\end{bmatrix}\begin{bmatrix} v \\ y\end{bmatrix} \geq \alpha|v|^2$$

$$\forall u,v \in R^n, \quad \forall x,y \in R^\ell, \quad \forall h \in H, \tag{3.1.5}$$

(C2) the functions $f(.,.,.)$, $D_u f(.,.,.)$ and $D_x f(.,.,.)$ are continuously differentiable in all variables,

(C3) the matrix functions $A(.)$ and $B(.)$ are continuously differentiable on H,

(C4) for each $h \in H$, $\phi^i(.,h)$, $i \in I$, are two times continuously differentiable and convex functions, while $\theta^j(.,h)$, $j \in J$, are three times continuously differentiable and convex functions,

(C5) the functions $\phi^i(.,.)$ and $D_u\phi^i(.,.)$, $i \in I$, are continuously differentiable on $R^n \times H$. Similarly $\theta^j(.,.)$, $D_x\theta^j(.,.)$ and $D^2_{xx}\theta^j(.,.)$, $j \in J$, are continuously differentiable on $R^\ell \times H$,

(C6) for each $h \in H$

$$\theta^j(x^o,h) < 0 \qquad \forall j \in J,$$

(C7) for each $h \in H$ there exist a pair $(\hat{u}_h,\hat{x}_h) \in L^\infty(0,T) \times C(0,T)$ which satisfies (3.1.2) and a constant $\rho(h) < 0$ such that

$$\phi^i(\hat{u}_h(t),h) \leq \rho(h) \quad \text{for all } i \in I \text{ and a.a. } t \in [0,T], \tag{3.1.6a}$$

$$\theta^j(\hat{x}_h(t),h) \leq \rho(h) \quad \text{for all } i \in I \text{ and all } t \in [0,T]. \tag{3.1.6b}$$

It is well known (see e.g. [33]) that under conditions (C1), (C4) and (C7) Problem (OC_h) has a unique solution.

For any $h \in H$ and $t \in [0,T]$ let us introduce the set of indices

$$I_h(t) = \{i \in I \mid \phi^i(u_h(t),h) = 0\}, \tag{3.1.7a}$$

$$J_h(t) = \{j \in J \mid \theta^j(x_h(t),h) = 0\} \tag{3.1.7b}$$

corresponding to all control and state constrants binding at t, for the optimal control and state respectively.

In addition to (C1) through (C7) we assume that for all $h \in H$ and almost all $t \in [0,T]$ the following constraint regularity condition similar to (B9) holds:

(C8) there exists a constant $\beta > 0$ such that

$$\left| \left[D_u \phi^T_{I_h(t)}(u_h(t),h), -B^T(h) D_x \theta^T_{J_h(t)}(x_h(t),h)\right] v \right| \geq \beta |v| \tag{3.1.8}$$

for almost every $t \in [0,T]$, every $h \in H$ and every v of appropriate dimension,
where $D_u \phi^T_{I_h(t)}(u_h(t),h)$ (respectively $D_x \theta^T_{J_h(t)}(x_h(t),h)$) denotes the matrix whose columns are the gradients of all constraints functions ϕ^i (resp. θ^j) binding at $u_h(t)$ (resp. $x_h(t)$).

Lemma 3.1

For any compact set $\mathcal{H} \subset H$ there exists a constant $c > 0$ such that

$$||u_h|| \leq c \qquad \forall h \in \mathcal{H}. \tag{3.1.9}$$

Proof

Let us take an arbitrary $h \in H$. From (C3), (C5) and (C7) it follows that there exists a constant $\varepsilon(h)$ such that for each $g \in S(h,\varepsilon(h))$ the pair $(\hat{u}_h, x_g(\hat{u}_h))$ satisfies

$$\dot{x}_g(\hat{u}_h)(t) = A(g) x_g(\hat{u}_h)(t) + B(g)\hat{u}_h(t), \quad x_g(\hat{u}_h)(0) = x^0 \tag{3.1.10}$$

and

$$\phi(\hat{u}_h(t),g) \leq \tfrac{1}{2}\rho(h), \quad \theta(x_g(\hat{u}_h)(t),g) \leq \tfrac{1}{2}\rho(h). \tag{3.1.11}$$

Hence $(\hat{u}_h, x_g(\hat{u}_h))$ is an admissible pair for (OC_g) and

$$F(u_g, x_g, g) \leq F(\hat{u}_h, x_g(\hat{u}_h), g). \tag{3.1.12}$$

Expanding $F(\cdot,\cdot,g)$ into Taylor's series at $(\hat{u}_h, x_g(\hat{u}_h), g)$ and taking into account (3.1.5) we get

$$\begin{aligned} F(u_g,x_g,g) \geq {} & F(\hat{u}_h, x_g(\hat{u}_h), g) + (D_u F(\hat{u}_h, x_g(\hat{u}_h), g), u_g - \hat{u}_h) + \\ & + (D_x F(\hat{u}_h, x_g(\hat{u}_h), g), x_g - x_g(\hat{u}_h)) + \alpha ||u_g - \hat{u}_h||^2. \end{aligned} \tag{3.1.13}$$

From (3.1.12) and (3.1.13) we have

$$||u_g - \hat{u}_h||^2 \leq \frac{1}{\alpha}\left[||D_u F(\hat{u}_h, x_g(\hat{u}_h), g)|| \, ||u_g - \hat{u}_h|| + ||D_x F(\hat{u}_h, x_g(\hat{u}_h), g)|| \, ||x_g - x_g(\hat{u}_h)||\right]. \tag{3.1.14}$$

On the other hand it follows from (C3) that

$$||x_g(\hat{u}_h)|| \leq c, \quad ||x_g - x_g(\hat{u}_h)|| \leq c\,||u_g - \hat{u}_h|| \qquad \forall g \in S(h,\varepsilon(h)).$$

Hence from (3.1.14) we obtain

$$||u_g|| \leq ||u_g - \hat{u}_h|| + ||\hat{u}_h|| \leq c \qquad \forall g \in S(h,\varepsilon(h)). \tag{3.1.15}$$

The set of the balls $S(h,\varepsilon(h))$ for all $h \in H$ constitutes an open covering of $\mathcal{H}$, from which we can extract a finite subcovering. This fact together with (3.1.15) imply (3.1.9).

□

3.2. Lagrange Formalism

In this section we are going to introduce Lagrange formalism for (OC_h). Before doing that we have to make few comments concerning the state space constraints (3.1.4).

Note that since x satisfying (3.1.2) is an absolutely continuous function, then by (C4) also $\theta(x(.),h)$ is an absolutely continuous function. Let

$$K^2 = \{y \in AC(0,T) \mid y(t) \geq 0 \quad \text{for all} \quad t \in [0,T]\} \tag{3.2.1}$$

be the cone of non-negative s-dimensional, absolutely continuous, vector functions.

Condition (3.1.4) can be expressed in the form

$$-\theta(x,h) \in K^2. \tag{3.2.2}$$

On the other hand since $AC(0,T) \subset C(0,T)$, then

$$K^2 \subset K^1 \overset{def}{=} \{y \in C(0,T) \mid y(t) \geq 0 \quad \text{for all} \quad t \in [0,T]\} \tag{3.2.3}$$

and condition (3.1.4) can as well be expressed as

$$-\theta(x,h) \in K^1. \tag{3.2.4}$$

For each representation (3.2.2) and (3.2.4) we obtain a little bit different forms of the Lagrangian associated with (OC_h).

We consider both cases, since each of them will be used in the sequel. We shall need the general form of linear continuous functionals non-negative on K^1 and K^2 respectively. The forms of these functionals depend on the topology introduced in the respectively spaces. We shall consider the following topologies

(1) space $C(0,T)$ with the uniform convergence topology i.e. supplied

with the norm

$$||y||_\infty = \max_{t\in[0,T]} \max_{1\leq i\leq s} |y^i(t)|.$$

In this topology the cone K^1 has interior points, and

$$\text{int } K^1 = \{y \in C(0,T) \,|\, y(t) > 0 \quad \text{for all} \quad t \in [0,T]\}. \tag{3.2.5}$$

Each linear continuous functional non-negative on K^1 can be expressed in the form of a Stieltjes integral

$$k(y) = \int_0^T \langle y(t), dk(t)\rangle \overset{def}{=} [k,y], \tag{3.2.6}$$

where k is a non-decreasing s-dimensional function of bounded variation, left-continuous on $[0,T)$ and vanishing at T:

$$k \in BV(0,T) \;; \quad k(T) = 0 \quad , \quad dk(t) \geq 0. \tag{3.2.7}$$

(2) space $AC(0,T)$ with the Sobolev topology $W^{1,1}(0,T)$, i.e. supplied with the norm

$$||y||_{1,1} = ||\dot{y}||_1 + |y(T)|. \tag{3.2.8}$$

Note that in this topology

$$\text{int } K^2 = \{y \in AC(0,T) \,|\, y(t) > 0 \quad \text{for all} \quad t \in [0,T]\}. \tag{3.2.9}$$

Each linear continuous functional defined on $W^{1,1}(0,T)$ can be expressed in the form

$$\ell(y) = -\int_0^T \langle y(t), \dot{\ell}(t)\rangle\, dt + \langle y(T), \ell(T)\rangle$$

where

$$\ell \in W^{1,\infty}(0,T),$$

$W^{1,\infty}(0,T)$ - is the Sobolev space supplied with the norm

$$||\ell||_{1,\infty} = ||\dot{\ell}||_\infty + |\ell(T)|$$

Since $\dot{\ell}$ and $\ell(T)$ are independent each of the other we shall use different symbols for each of these elements putting

$$\ell_1 = \dot{\ell} \quad , \quad \ell_2 = \ell(T).$$

Hence any linear continuous functional defined on $W^{1,1}(0,T)$ can be expressed in the form

$$\ell(y) = -\int_0^T <\dot{y}(t),\ \ell_1(t)> dt + <y(T),\ell_2> \tag{3.2.10}$$

where

$$\ell_1 \in L^\infty(0,T) \quad , \quad \ell_2 \in R^s. \tag{3.2.11}$$

After simple evaluation we get from (3.2.10)

$$\ell(y) = -\int_0^T <\dot{y}(t),\ \ell_1(t) - \ell_2> dt + <y(0),\ell_2>. \tag{3.2.12}$$

Analysis of (3.2.10) and (3.2.12) shows that any linear continuous functional non-negative on K_2 can be expressed in the form (3.2.10) where

$$\ell_1(t) \geq 0, \qquad \text{almost everywhere on } [0,T], \tag{3.2.13a}$$

$$\ell_2 - \ell_1(t) \geq 0, \quad \text{almost everywhere on } [0,T], \tag{3.2.13b}$$

$$\ell_1(\cdot) \text{ - is non-decreasing almost everywhere on } [0,T]. \tag{3.2.13c}$$

Note that for $y \in AC(0,T)$ functional $k(y)$ given by (3.2.6), (3.2.7) after intergation by parts can be rewritten as

$$k(y) = -\int_0^T <\dot{y}(t),\ k(t)> dt - < y(0),\ k(0)>. \tag{3.2.14}$$

Comparing (3.2.12) and (3.2.14) we find that each functional defined on $AC(0,T)$ by (3.2.6) can be expressed in the form (3.2.10), where

$$\ell_2 = -k(0),\ \ell_1(t) - k(t) - k(0) \quad \text{for almost all } t \in [0,T]. \tag{3.2.15}$$

Now we shall introduce a Lagrangian for (OC_h) using representation (1) of the constraints (3.1.4). In a classical way we get

$$L(.,.;.,.,.;.) : L^2(0,T) \times C(0,T) \times L^2(0,T) \times BV(0,T) \times H \to R^1,$$

$$L(u,x;p,\lambda,\nu;h) = F(u,x,h) + (p,\dot{x} - A(h)x - B(h)u) + (\lambda,\phi(u,h)) + [\nu,\theta(x,h)]. \tag{3.2.16}$$

Since (OC_h) is a convex problem and the Slater's conditions (C7) are satisfied the following result holds (see [25]):

<u>Lemma 3.2</u>

If conditions (C1) through (C7) are satisfied then there exist Lagrange multipliers $p_h, \lambda_h \in L^2(0,T)$ and $\nu_h \in BV(0,T)$ such that the unique solution (u_h,x_h) of (OC_h) is characterized by the saddle point of Lagrangian (3.2.16) i.e.

$$L(u_h,x_h;p,\lambda,\nu;h) \leq L(u_h,x_h;p_h,\lambda_h,\nu_h;h) \leq L(u,x;p_h,\lambda_h,\nu_h;h)$$

$\forall u \in L^2(0,T)$; $\forall x \in C(0,T)$, $x(0)=x^o$, $\forall p \in L^2(0,T)$; $\forall \lambda \in L^2(0,T)$, $\lambda(t) \geq 0$;

$$\forall \nu \in BV(0,T),\ \nu(T)=0,\ d\nu(t) \geq 0. \tag{3.2.17}$$

Note that (3.2.17) isequivalent to the following stationarity conditions

$$p_h(t)-\int_t^T [A^T(h)p_h(\tau)-D_x f(u_h(\tau),x_h(\tau),h)]d\tau +$$
$$+\int_t^T D_x\theta^T(x_h(\tau)),h)d\nu_h(\tau) = 0 \quad \text{for almost all} \quad t \in [0,T], \tag{3.2.18}$$

$$D_u f(u_h(t),x_h(t),h)-B^T(h)p_h(t)+D_u\phi^T(u_h(t),h)\lambda_h(t) = 0$$
$$\text{for almost all} \quad t \in [0,T], \tag{3.2.19}$$

along with complementary slackness

$$(\lambda_h,\ \phi(u_h,h)) = 0, \qquad \lambda_h(t) \geq 0, \tag{3.2.20}$$

$$[\nu_h,\ \theta(x_h,h)] = 0, \qquad d\nu_h(t) \geq 0. \tag{3.2.21}$$

Note that (C6) together with (3.2.21) imply

$$\nu_h(0) = \nu_h(0^+). \tag{3.2.22}$$

Following W.W. Hager [24] it is convenient to introduce, instead of p_h, a new dual variable q_h defined by

$$q_h(t) = D_x\theta^T(x_h(t),h)\nu_h(t)-p_h(t). \tag{3.2.23}$$

Substituting (3.2.23) into (3.2.18) we find that q_h satisfies the following differential equation

$$\dot{q}_h(t) = -A^T(h)q_h(t)+[A^T(h)D_x\theta^T(x_h(t),h)+D^2_{xx}\theta^T(x_h(t),h)\dot{x}_h(t)]\nu_h(t)+$$
$$-D_x f(u_h(t),x_h(t),h), \tag{3.2.24}$$

$$q_h(T) = 0. \tag{3.2.24a}$$

Hence q_h is an absolutely continuous function.
Condition (3.2.19) expressed in terms of q_h takes on the form

$$D_u f(u_h(t),x_h(t),h)+B^T(h)q_h(t)-B^T(h)D_x\theta^T(x_h(t),h)\nu_h(t) +$$
$$+ D_u\phi^T(u_h(t),h)\lambda_h(t)=0 \quad \text{for almost all} \quad t \in [0,T]. \tag{3.2.25}$$

We introduce still another Lagrangian associated with (OC_h) using representation (2) of constraints (3.1.4). Namely

$$L_2(.,.;.,.,.,.;.) : L^2(0,T) \times H^{1,1}(0,T) \times L^2(0,T) \times L^\infty(0,T) \times R^s \times H \to R^1,$$

$$L^2(u,x;p,\lambda,\pi,\sigma;h) = F(u,x,h)+(p,\dot{x}-A(h)x-B(h)u)+(\lambda,\phi(u,h))+(\pi,D_x\theta(x,h)\overset{\sigma}{x})+$$
$$-<\sigma,\theta(x(T),h)>. \tag{3.2.26}$$

For L_2 the saddle point condition analogous to (3.2.16) holds:

$$L_2(u_h,x_h;p,\lambda,\pi,\sigma;h)\leqslant L_2(u_h,x_h;p_h,\lambda_h,\pi_h,\sigma_h;h) \leqslant L(u,x;p_h,\lambda_h,\pi_h,\sigma_h;h)$$

$$\forall u\in L^2(0,T);\ \forall x\in W^{1,1}(0,T),\ x(0)=x^o;\ \forall p\in L^2(0,T);\ \forall\lambda\in L^2(0,T),\ \lambda(t)\geqslant 0;$$

$$\forall\pi\in L^\infty(0,T),\ \pi(t)\geqslant 0,\ \pi(t) - \text{non-decreasing};\ \forall\sigma\in R^s, \sigma-\pi(T)\geqslant 0. \tag{3.2.27}$$

It is equivalent to

$$p_h(t)-\int_t^T [A^T(h)p_h(\tau)-D_xf(u_h(\tau),x_h(\tau),h)+(D^2_{xx}\theta^T(x_h(\tau),h)\dot{x}_h(\tau))\pi_h(\tau)]d\tau +$$
$$-D_x\theta^T(x_h(t),h)\pi_h(t)+D_x\theta^T(x_h(T),h)\sigma_h =$$
$$=p_h(t)-\int_t^T [A^T(h)p_h(\tau)-D_xf(u_h(\tau),x_h(\tau),h)+ \frac{d}{d\tau}(D_x\theta^T(x_h(\tau),h))\pi_h(\tau)]d\tau +$$
$$-D_x\theta^T(x_h(t),h)\pi_h(t)+D_x\theta^T(x_h(T),h)\sigma_h=0 \quad \text{for a.a. } t\in[0,T], \tag{3.2.28}$$

$$D_uf(u_h(t),x_h(t),h)-B^T(h)p_h(t)+D_u\phi^T(u_h(t),h)\lambda_h(t) = 0$$
$$\text{for almost all } t\in[0,T], \tag{3.2.29}$$

$$(\lambda_h,\phi(u_h,h)) = 0, \tag{3.2.30}$$

$$-(\pi_h,D_x\theta(x_h,h)\dot{x})+<\sigma_h,\theta(x_h(T),h)>= 0, \tag{3.2.31}$$

where according to (3.2.13)

$$\pi_h^j(t)\geqslant 0, \tag{3.2.32a}$$

$$\pi_h^j \text{ - is non-decreasing almost everywhere on } [0,T], \tag{3.2.32b}$$

$$\sigma_h^j-\pi_h^j(t)\geqslant 0 \quad \text{for almost all} \quad t\in[0,T]. \tag{3.2.32c}$$

By (3.2.15) we have

$$\sigma_h = -\nu_h(0), \quad \pi_h(t)=\nu_h(t)-\nu_h(0) \quad \text{for almost all} \quad t \in [0,T]. \tag{3.2.33}$$

Note that (3.2.22) and (3.2.32) imply

$$\operatorname*{ess\,inf}_{t \in [0,T]} \pi_h^j(t) = 0. \tag{3.2.34}$$

It is easy to see that conditions (3.2.31), (3.2.32) and (3.2.34) are satisfied if and only if

π_h^j - is non-decreasing almost everywhere on $[0,T]$

and for any subinterval $M \subset [0,T]$ such that

$$\theta^j(x_h(t),h) < 0 \quad \forall t \in M \quad \text{we have} \quad \pi_h^j = \text{const on } M, \tag{3.2.35a}$$

$$\operatorname*{ess\,inf}_{t \in [0,T]} \pi_h(t) = 0, \tag{3.2.35b}$$

$$\sigma_h^j \geq \operatorname*{ess\,sup}_{t \in [0,T]} \pi_h^j(t) \quad \text{and}$$

$$\sigma_h^j = \operatorname*{ess\,sup}_{t \in [0,T]} \pi_h(t) \quad \text{if} \quad \theta^j(x_h(T),h) < 0. \tag{3.2.35c}$$

In the case where $\dot{\pi} \in L^2(0,T)$ we can intergate by parts the last term under the integral in (3.2.28) to obtain

$$p_h(t) - \int_t^T [A^T(h)p_h(\tau) - D_x f(u_h(\tau),x_h(\tau),h) - D_x\theta^T(x_h(\tau),h)\dot{\pi}_h(\tau)]d\tau +$$

$$+ D_x\theta^T(x_h(T),h)(\sigma_h - \pi_h(T)) = 0. \tag{3.2.36}$$

3.3. Regularity of Primal and Dual Variables

Regularity of primal and dual optimal variables for (OC_h) was investigated by W.W. Hager [24]. In this section we shall present both the results and the proofs based on [24]. Like in that paper we shall use the form (3.2.24), (3.2.25) of optimality conditions for (OC_h), to obtain Lipschitz continuity of $u_h(.)$, $p_h(.)$, $\lambda_h(.)$ and $\nu_h(.)$. To this end we shall need several auxiliary lemmas.

Lemma 3.3

After possible correction on a set of measure zero, u_h has bounded variation and is left-continuous on $[0,T)$.

Proof

Functions $u_h(.)$ and $\lambda_h(.)$ are defined almost everywhere on $[0,T]$, but we can (see [24] Appendix B) redefine them on a set of measure zero in such a way that (3.2.25) is satisfied for all $t \in [0,T]$.

Condition (3.2.25) together with (3.2.20) implies that for any $t \in [0,T]$ $u_h(t)$ can be treated as the solution of the following convex programming problem

$$\min_{u \in \Phi_h} \{d(u,z_h(t),h) \overset{\text{def}}{=} f(u,x_h(t),h)+\langle B^T(h)\left[q_h(t)-D_x\theta^T(x_h(t),h)\nu_h(t)\right],u\rangle\} \tag{3.3.1}$$

which for a fixed h depends on the vector $z_h^T(t)=(x_h^T(t),q_h^T(t),\nu_h^T(t))$.

We shall show that the solutions of (3.3.1) are Lipschitz continuous functions of $z_h(t)$, i.e. that

$$|u_h(t_2)-u_h(t_1)| \leq c|z_h(t_2)-z_h(t_1)| \qquad \forall t_1,t_2 \in [0,T]. \tag{3.3.2}$$

Indeed using optimality conditions for (3.3.1) at the points t_1 and t_2, in the form (1.1.5), we get

$$\langle D_u d(u_h(t_1),z_h(t_1),h)-D_u d(u_h(t_2),z_h(t_2),h),u_h(t_2)-u_h(t_1)\rangle \geq 0.$$

Therefore

$$\langle D_u d(u_h(t_1),z_h(t_1),h)-D_u d(u_h(t_1),z_h(t_2),h),u_h(t_2)-u_h(t_1)\rangle \geq$$

$$\geq \langle D_u d(u_h(t_1),z_h(t_2),h)-D_u d(u_h(t_2),z_h(t_2),h),u_h(t_1)-u_h(t_2)\rangle.$$

Since by (3.3.1)

$$D_u d(u_h(t),z_h(t),h)=D_u f(u_h(t),x_h(t),h)+B^T(h)q_h(t)-B^T(h)D_x\theta^T(x_h(t),h)\nu_h(t),$$

then taking into account (C1) and (C4) as well as the fact that $x_h(t)$ is bounded uniformly on $[0,T]$ we obtain

$$\alpha|u_h(t_2)-u_h(t_1)|^2 \leq c|z_h(t_2)-z_h(t_1)|\,|u_h(t_2)-u_h(t_1)|,$$

which implies (3.3.2).

Since $z_h(.)$ is a left-continuous function of bounded variation (3.3.2) completes the proof of the lemma □

Corollary 3.1

$\overset{\circ}{x}_h$ and $\dot{q}_h$ are left-continuous functions of bounded variation on $[0,T]$.

Proof

Since in Lemma 3.3 we have redefined functions u_h and λ_h to the whole interval (0,T) the equations (3.1.2) and (3.2.24) must be treated as being satisfied everywhere on (0,T). These equations together with Lemma 3.2 imply the corollary. □

Let g be a left-continuous function of bounded variation defined on (0,T), then for each $\sigma \in (0,T)$ there exists the right limit

$$g(\sigma^+) \overset{\text{def}}{=} \lim_{t \downarrow \sigma} g(t).$$

Lemma 3.4

If for some $\sigma \in (0,T)$

$$\theta^j(x_h(\sigma),h) = 0, \tag{3.3.3}$$

then

$$D_x\theta^j(x_h(\sigma),h)\left[A(h)x_h(\sigma)+B(h)u_h(\sigma)\right] \geq 0, \tag{3.3.4a}$$

$$D_x\theta^j(x_h(\sigma),h)\left[A(h)x_h(\sigma)+B(h)u_h(\sigma^+)\right] \leq 0, \tag{3.3.4b}$$

hence

$$D_x\theta^j(x_h(\sigma),h)B(h)\left[u_h(\sigma^+)-u_h(\sigma)\right] \leq 0. \tag{3.3.5}$$

Proof

Since $\dot{x}_h$ has a bounded variation there exists the right-limit $D_x\theta^j(x_h(\sigma),h)\dot{x}_h(\sigma^+)=\lim_{t\downarrow\sigma} D_x\theta^j(x_h(t),h)\dot{x}_h(t)$. If $D_x\theta^j(x_h(\sigma),h)x_h(\sigma^+) \leq 0$ then taking into account (3.1.2) we get (3.3.4b).

Now let us suppose that there exists an interval $[\sigma,\sigma+\Delta]$ such that $D_x\theta^j(x_h(t),h)\dot{x}_h(t) > 0$ for all $t \in (\sigma,\sigma+\Delta)$.

Hence, taking into acount that $\theta^j(x_h(\sigma+\Delta),h) \leq 0$, we should have

$$0=\theta^j(x_h(\sigma),h)=\theta^j(x_h(\sigma+\Delta))-\int_\sigma^{\sigma+\Delta} D_x\theta^j(x_h(t),h)\dot{x}_h(t)dt < 0.$$

It follows from the obtained contradiction that there exists a sequence $\{t_k\}\downarrow\sigma$ such that $D_x\theta^j(x(t_k),h)\dot{x}_h(t_k) \leq 0$, which implies (3.3.4b). In exactly the same way we obtain (3.3.4a). Combination of these relations yields (3.3.5). □

Lemma 3.5

Optimal control u_h is continuous and uniformly bounded on (0,T).

Proof

Uniform boundedness follows from the fact that u_h has bounded variation on $(0,T)$. Since by Lemma 3.3 u_h is a left-continuous function it is enough to show that it is right-continuous.

Note that by (C4) and by (3.3.2)

$$\phi(u_h(\sigma^+),h) \leqslant 0. \qquad (3.3.6)$$

Hence by (3.2.20), (3.2.25) and (3.3.2) $u_h(\sigma^+)$ is the solution of (3.3.1) with $z_h(t)$ substituted by $z_h(\sigma^+)$.

Since convex programming problem (3.3.1) has a unique solution we have

$$d(u_h(\sigma^+),z_h(\sigma^+),h) < d(u_h(\sigma),z_h(\sigma^+),h) \quad \text{if} \quad u_h(\sigma^+) \neq u_h(\sigma). \qquad (3.3.7)$$

It will be shown that $d(u_h(\sigma^+),z_h(\sigma^+),h) \geqslant d(u_h(\sigma),z_h(\sigma^+),h)$ which implies $u_h(\sigma^+)=u_h(\sigma)$, i.e. the continuity of u_h.

Indeed, recalling definition $d(u,z,h)$ given in (3.3.1) we get

$$\langle D_u d(u_h(\sigma),z_h(\sigma^+),h),u_h(\sigma^+)-u_h(\sigma)\rangle = \langle D_u d(u_h(\sigma),z_h(\sigma),h),u_h(\sigma^+)-u_h(\sigma)\rangle +$$

$$- \langle v_h(\sigma^+)-v_h(\sigma),D_x\theta(x_h(\sigma),h)B(h)(u_h(\sigma^+)-u_h(\sigma))\rangle. \qquad (3.3.8)$$

Since

$$v_h^j(\sigma^+)-v_h^j(\sigma) \geqslant 0$$

and by (3.2.21) it may happen that

$$v_h^j(\sigma^+)-v_h^j(\sigma) > 0 \quad \text{only if} \quad \theta^j(x_h(\sigma),h)=0,$$

then from (3.3.5) and (3.3.8) we obtain

$$\langle D_u d(u_h(\sigma),z_h(\sigma^+),h),u_h(\sigma^+)-u_h(\sigma)\rangle \geqslant \langle D_u d(u_h(\sigma),z_h(\sigma),h),u_h(\sigma^+)-u_h(\sigma)\rangle \geqslant 0, \qquad (3.3.9)$$

where the last inequality follows from optimality condition of the form (1.1.5) for (3.3.1).

Convexity of $d(\cdot,z_h(\sigma^+),h)$ together with (3.3.9) yield

$$d(u_h(\sigma),z_h(\sigma^+),h) \leqslant d(u_h(\sigma^+),z_h(\sigma^+),h),$$

which imply $u_h(\sigma)=u_h(\sigma^+)$. □

From Lemma 3.4 and Lemma 3.5 we get

Corollary 3.2

For every $t \in (0,T)$ such that

$$\theta^j(x_h(t),h)=0 \tag{3.3.10}$$

we have

$$D_x\theta^j(x_h(t),h)[A(h)x_h(t)+B(h)u_h(t)] = 0. \tag{3.3.11}$$

Lemma 3.6

Functions λ_h and ν_h are continuous and uniformly bounded on $(0,T)$.

Proof

Uniform boundedness of ν_h follows from the fact that it has bounded variation. To show uniform boundedness of λ_h note that from (3.2.25) we get

$$|D_u\phi^T(u_h(t),h)\lambda_h(t)| \leq |D_uf(u_h(t),x_h(t),h)|+|B^T(h)q_h(t)| +$$

$$+ |B^T(h)D_x\theta^T(x_h(t),h)\nu_h(t)|. \tag{3.3.12}$$

Let us denote by $\lambda_{h,I_h(t)}$ and $\nu_{h,J_h(t)}$ the subvectors of λ_h and ν_h containing all components $i \in I_h(t)$ and $j \in J_h(t)$ respective i.e. corresponding to the constraints binding at t.

Since by (3.2.20) and by continuity of $u_h(.)$ $\lambda_h^i(t) \neq 0$ only if $i \in I_h(t)$, then from (3.1.8) and (3.3.12) we obtain

$$|\lambda_h(t)|=|\lambda_{h,I_h(t)}(t)| \leq \frac{1}{\beta} |D_u\phi^T_{I_h(t)}(u_h(t),h)\lambda_{h,I_h(t)}(t)| =$$

$$= \frac{1}{\beta} |D_u\phi^T(u_h(t),h)\lambda_h(t)| \leq \frac{1}{\beta}\{|D_uf(u_h(t),x_h(t),h)|+|B^T(h)q_h(t)| +$$

$$+|B^T(h)D_x\theta^T(x_h(t),h)\nu_h(t)|\},$$

which together with the uniform boundedness of $u_h(t)$ and $\nu_h(t)$ shows that $\lambda_h(t)$ is uniformly bounded.

Now let us prove continuity.

Let $\sigma \in (0,T)$ be an arbitrary point. Since λ_h is uniformly bounded there exists a sequence $\{t_k\} \subset (0,T)$ and $\mu \in R^r$ such that

$$t_k \downarrow \sigma \quad \text{and} \quad \lambda(t_k) \to \mu$$

Condition of optimality (3.2.25) at the points t_k yields

$$D_uf(u_h(t_k),x_h(t_k),h)+B^T(h)q_h(t_k)-B^T(h)D_x\theta^T(x_h(t_k),h)\nu_h(t_k) +$$

$$+ D_u\phi^T(u_h(t),h)\nu_h(t) = 0$$

Passing to the limit and using Lemma 3.5 we obtain

$$D_u f(u_h(\sigma),x_h(\sigma),h)+B^T(h)q_h(\sigma)-B^T(h)D_x\theta^T(x_h(\sigma),h)\nu_h(\sigma^+) +$$

$$+D_u\phi^T(u_h(\sigma),h)\mu = 0$$

Combining this equation with condition of optimality (3.2.25) at σ we get

$$-B^T(h)D_x\theta^T(x(\sigma),h)\left[\nu_h(\sigma^+)-\nu_h(\sigma)\right]+D_u\phi^T(u_h(\sigma),h)\left[\mu-\lambda_h(\sigma)\right]=0 \quad (3.3.13)$$

By (3.2.21) $\nu_h^j(\sigma^+)\neq\nu_h^j(\sigma)$ only if $\theta^j(x_h(\sigma),h)=0$, similarly by continuity of $u_h(.)$ and by (3.2.20) $\mu^i\neq\lambda^i(\sigma)\neq 0$ only if $\phi^i(u_h(\sigma),\sigma)=0$. Hence we can rewrite (3.3.13) in the form

$$-B^T(h)D_x\theta^T_{J_h(\sigma)}(x_h(\sigma),h)\left[\nu_{h,J_h(\sigma)}(\sigma^+)-\nu_{h,J_h(\sigma)}(\sigma)\right] +$$

$$+ D_u\phi^T_{I_h(\sigma)}\left[\mu_{I_h(\sigma)}-\lambda_{h,I_h(\sigma)}(\sigma)\right] = 0. \quad (3.3.14)$$

Assumption (C8) together with (3.3.14) imply

$$\nu_{h,J_h(\sigma)}(\sigma^+)-\nu_{h,J_h(\sigma)}(\sigma) = 0 \ , \quad \mu_{I_h(\sigma)} = \lambda_{h,I_h(\sigma)}(\sigma),$$

hence also

$$\nu_h(\sigma^+)-\nu_h(\sigma) = 0 \qquad , \quad \mu-\lambda_h(\sigma) = 0.$$

Thus we proved right-continuity of ν_h and λ_h. Repeating the same argument for a sequence convergent to σ from the left we conclude the proof of the lemma.

□

From Lemmas 3.5 and 3.6 as well as from (3.1.2) and (3.2.24) we obtain

Corollary 3.3

Fuctions $\ddot{x}_h$ and $\dot{q}_h$ are continuous and uniformly bounded on $(0,T)$.

Now following W.W. Hager it will be shown that u_h, λ_h and ν_h are Lipschitz continuous functions on $[0,T)$. To this purpose Theorem 1.2 will be used.

To apply Theorem 1.2 we put

$H = (0,T) \subset R^1$

$Y = R^{n+r+s}$

$k^T(t) = [u_h^T(t), \lambda_h^T(t), \nu_h^T(t)]$

$I(t) = I_h(t) \cup J_h(t)$ = subset of the set $\{1,2,\dots,r+s\}$ containing all indices of the state and control constraints binding at t.

By Lemmas 3.5 and 3.6 $k(.)$ is a continuous function, while by continuity of u_h and x_h as well as by (C4) the upper semicontinuity property (1.3.1) is satisfied by $I(.)$. Hence all assumptions of Theorem 1.2 hold.

In our case a compatible pair (τ,σ) is any pair of points $\tau,\sigma \in (0,T)$ such that at τ and σ the same constraints are binding, while in the segment $[\tau,\sigma]$ none other constraint is active.

Below it will be shown that there exist constants $\gamma < \infty$ and $\eta > 0$ such that for any compatible pair (τ,σ) satisfying

$$|\tau - \sigma| \leq \eta \tag{3.3.15}$$

we have

$$|k(\tau)-k(\sigma)| \leq \gamma\, |\tau-\sigma|. \tag{3.3.16}$$

By Theorem 2.1 this result will imply that (3.3.16) holds for any pair $\tau,\sigma \in (0,T)$ i.e. u_h, λ_h and ν_h are Lipschitz continuous on $(0,T)$.

Let (τ,σ) be any arbitrary compatible pair, i.e.

$$I_h(\tau) = I_h(\sigma) \overset{\text{def}}{=} I_h \quad\text{and}\quad J_h(\tau) = J_h(\sigma) \overset{\text{def}}{=} J_h.$$

Moreover let us denote

$$K_h(\tau)=K_h(\sigma) \overset{\text{def}}{=} K_h=J\setminus J_h(\tau)=\{j \in J \,|\, \theta^j(x_h(\tau),h)<0\}.$$

By θ_{K_h} we denote the subvector of θ containing all components whose indices belong to K_h. Corresponding subvector of ν_h will be denoted by ν_{h,K_h}.

From (3.1.7a), (3.2.25) and (3.3.11) it follows that at the points $t=\tau$ and $t=\sigma$ the following system of equations holds

$$D_u f(u_h(t),x_h(t),h)+D_u\phi_{I_h}^T(u_h(t),h)\lambda_{h,I_h}(t)-B^T(h)D_x\theta_{J_h}^T(x_h(t),h)\nu_{h,J_h}(t)+$$

$$-B^T(h)D_x\theta_{K_h}^T(x_h(t),h)\nu_{h,K_h}(t)+B^T(h)q_h(t) = 0, \tag{3.3.17a}$$

$$D_u\phi_{I_h}(u_h(t),h) = 0, \tag{3.3.17b}$$

$$-D_x\theta_{J_h}(x_h(t),h)B(h)u_h(t)-D_x\theta_{J_h}(x_h(t),h)A(h)x_h(t) = 0. \tag{3.3.17c}$$

To simplify notation let us put

$$w^T(t)\overset{def}{=}[u_h^T(t),\lambda_{h,I_h}^T(t),\nu_{h,J_h}^T(t)],\ \ell^T(t)=[x_h^T(t),q_h^T(t),\nu_{h,K_h}^T(t)] \tag{3.3.18}$$

and write (3.3.17) as

$$G(w(t),\ell(t))=0. \tag{3.3.19}$$

In this equation we shall treat $w(t)$ as an unknown and $k(t)$ as a parameter.

We introduce an auxiliary equation depending on a scalar parameter δ putting

$$\overline{G}(\overline{w},\delta) \overset{def}{=} G(\overline{w},\overline{\ell}(\delta))=0 \qquad 0 \leq \delta \leq 1, \tag{3.3.20}$$

where

$$\overline{\ell}(\delta) \overset{def}{=} (1-\delta)\ell(\tau)+\delta\ell(\sigma). \tag{3.3.20a}$$

In the way analogous to that in Section 1.4, using implicit function theorem, we shall show that for $\delta \in [0,1]$ the solution $\overline{w}(\delta)$ of (3.3.20) exists,is unique and moreover it is a differentiable function of δ on (0,1), provided that in (3.3.15) η is sufficiently small.

Since at $\delta=0$ and $\delta=1$ equation (3.3.20) coincides with (3.3.19) we have

$$\overline{w}(0) = w(\tau) \ , \ \overline{w}(1) = w(\sigma), \tag{3.3.21}$$

and we can derive the needed properties of $w(.)$ from the properties of $\overline{w}(.)$.

From (3.3.17) and (3.3.20) we get

$$D_u\overline{G}(\overline{w},\delta)=\begin{bmatrix} D^2_{uu}f(\overline{u},\overline{x}(\sigma),h)+D^2_{uu}\phi^T_{I_h}(\overline{u},h)\overline{\lambda}_{I_h}, & D_u\phi^T_{I_h}(\overline{u},h), & -B^T(h)D_x\theta^T_{J_h}(\overline{x}(\delta),h) \\ D_u\phi_{I_h}(\overline{u},h) & 0 & 0 \\ -D_x\theta_{J_h}(\overline{x}(\delta),h)B(h) & 0 & 0 \end{bmatrix} \tag{3.3.22}$$

Note that the matrix $D_w\overline{G}(\overline{w},\delta)$ has the same structure as the matrix A given by (1.4.8). Using this fact, as well as (C1) and (C8) and repeating the argument of the proof of Lemma 2.2 we find that if

$$|x_h(\sigma)-x_h(\tau)| \leq \xi, \qquad (3.3.23a)$$

$$|\bar{w}- w(\tau)| \leq \zeta, \qquad (3.3.23b)$$

where $\xi > 0$ and $\zeta > 0$ are some constants independent of $\tau \in (0,T)$, then $D_w\bar{G}(\bar{w},\delta)$ is nonsingular and

$$|[D_w\bar{G}(\bar{w},\delta)]^{-1}| \leq \max\{ \frac{2}{\alpha}, \frac{1}{\beta} [(B^2(\delta))+2\beta)^{\frac{1}{2}} + B(\delta)]\} \qquad (3.3.24)$$

where

$$B(\delta) = |D^2_{uu}f(\bar{u},\bar{x}(\delta),h)+D^2_{uu}\phi_{I_h}(\bar{u},h)\bar{\lambda}_{I_h}|. \qquad (3.3.24a)$$

Note that by Lemmas 3.5 and 3.6 as well as by (3.3.23) $\bar{u},\bar{\lambda}_h$ and $\bar{x}(\delta)$ are bounded uniformly with respect to $\tau \in (0,T)$. Hence from (3.3.24) we find that there exists a constant $c_o > 0$ independent of $\tau \in (0,T)$ such that

$$|[D_w\bar{G}(\bar{w},\delta)]^{-1}| \leq c_o. \qquad (3.3.25)$$

Thus, by implicit function theorem there exists a continuous function $\bar{w}(\delta)$ such that

$$\bar{G}(\bar{w}(\delta),\delta) = 0.$$

Note that by the definition of a compatible pair

$$\lambda^i_h(\tau) = \lambda^i_h(\sigma) = 0 \quad \text{for} \quad i \notin I_h \qquad (3.3.26a)$$

and

$$\nu^j_h(\tau) - \nu^j_h(\sigma) = 0 \quad \text{for} \quad j \notin J_h. \qquad (3.3.26b)$$

Hence from (3.3.17) and (3.3.20) we obtain

$$D_\delta\bar{G}(w,\delta) = \begin{bmatrix} z_{1,1}(\delta)+z_{1,2}(\delta)\bar{\nu}_{J_h} \\ 0 \\ z_{3,1}(\delta)+z_{3,2}(\delta)\bar{u} \end{bmatrix}, \qquad (3.3.27)$$

where

$$z_{1,1}(\delta)=D^2_{ux}f(\bar{u},\bar{x}_h(\delta),h)(x_h(\sigma)-x_h(\tau))+B^T(h)(q_h(\sigma)-q_h(\tau)) +$$

$$-B^T(h)D^2_{xx}\theta^T_{K_h}(\bar{x}_h(\delta),h)(x_h(\sigma)-x_h(\tau))\nu_{h,K_h}(\sigma), \qquad (3.3.27a)$$

$$z_{1,2}(\delta)=-B^T(h)D^2_{xx}\theta^T_{J_h}(\bar{x}_h(\delta),h)(x_h(\sigma)-x_h(\tau)), \qquad (3.3.27b)$$

$$z_{3,1}(\delta)=-D^2_{xx}\theta_{J_h}(\bar{x}_h(\delta),h)(x_h(\sigma)-x_h(\tau))A(h)\bar{x}_h(\delta) -$$

$$-D_x\theta_{J_h}(\bar{x}_h(\delta),h)A(h)(x_h(\sigma)-x_h(\tau)), \qquad (3.3.27c)$$

$$z_{3,2}(\delta)=-D^2_{xx}\theta_{J_h}(\bar{x}_h(\delta),h)(x_h(\sigma)-x_h(\tau))B(h). \qquad (3.3.27d)$$

By implicit function theorem it follows that

$$D_\delta\bar{w}(\delta) = -[D_w\bar{G}(\bar{w},\delta)]^{-1}[D_\delta\bar{G}(\bar{w},\delta)].$$

Hence

$$|D_\delta\bar{w}(\delta)| \leq |[D_w\bar{G}(\bar{w},\delta)]^{-1}||D_\delta\bar{G}(\bar{w},\delta)|. \qquad (3.3.28)$$

Taking into account (3.3.23) and Lemma 3.5 we obtain from (3.3.27)

$$|D_\delta\bar{G}(\bar{w},\delta)| \leq c_1(|x_h(\sigma)-x_h(\tau)|+|q_h(\sigma)-q_h(\tau)|)+c_2|\bar{w}(\delta)||x_h(\sigma)-x_h(\tau)|. \qquad (3.3.29)$$

It follows from (3.3.25), (3.3.28) and (3.3.29) that if conditions (3.3.23) hold then

$$|D_\delta\bar{w}(\delta)| \leq c_3(|x_h(\sigma)-x_h(\tau)|+|q_h(\sigma)-q_h(\tau)|)+$$

$$+ c_4|\bar{w}(\delta)||x_h(\sigma)-x_h(\tau)|.$$

Hence taking into account that by (3.3.18), (3.3.21) and by Lemmas 3.5 and 3.6 $\bar{w}(0)=w(\tau)$ is uniformly bounded on $(0,T)$ and using Gronwall's lemma [23] we get

$$|D_\delta\bar{w}(\delta)| \leq c_5(|x_h(\sigma)-x_h(\tau)|+|q_h(\sigma)-q_h(\tau)|), \qquad (3.3.30)$$

where c_5 does not depend on $\tau,\sigma \in (0,T)$.
From (3.3.21) and (3.3.30) we have

$$|w(\sigma)-w(\tau)|=|\bar{w}(1)-\bar{w}(0)| \leq \int_0^1 |D_\delta\bar{w}(\delta)|d\delta \leq c_5(|x_h(\sigma)-x_h(\tau)|+|q_h(\sigma)-q_h(\tau)|). \qquad (3.3.31)$$

Since by Corollary 3.3 functions x_h and q_h are Lipschitz continuous on $(0,T)$, then from (3.3.31) we eventually get

$$|w(\sigma)-w(\tau)| \leq \gamma|\sigma-\tau|, \qquad (3.3.32)$$

where γ does not depend on τ and σ.
(3.3.32) together with (3.3.26) imply (3.3.16).

Still the assumptions (3.3.23) have to be verified.
From (3.3.31) it follows that (3.3.23b) is satified if

$$|x_h(\sigma)-x_h(\tau)|+|q_h(\sigma)-q_h(\tau)| \leq \frac{\xi}{c_5} \ . \qquad (3.3.33)$$

Since x_h and q_h are Lipschitz continuous, there exists a constant $\eta > 0$ independent of $\tau,\sigma \in (0,T)$ such that (3.3.15) implies (3.3.23a) and (3.3.33).

In this way it was shown that (3.3.16) is satisfied for each compatible pair satisfying (3.3.15).

Using Theorem 2.1 we get:

Theorem 3.1

If conditions (C1) through (C8) hold then the optimal control u_h and the optimal Lagrange multipliers λ_h and ν_h are Lipschitz continuous functions on $(0,T)$.

Note that by (C6) and by (3.2.19) we have

$$\nu_h(0) = \lim_{t \downarrow 0} \nu_h(t) \qquad (3.3.34)$$

On the other hand the function λ_h, which satisfies homogeneous terminal condition $\nu_h(T)=0$, in general is not continuous at $t=T$. It is obvious that we do not destroy optimality putting

$$u_h(0) = \lim_{t \downarrow 0} u_h(t) \ , \quad \lambda_h(0) = \lim_{t \downarrow 0} \lambda_h(t)$$

Thus from Theorem 3.1 as well as from (3.1.2), (3.2.23) and (3.2.24) we obtain

Corollary 3.4

Functions $u_h, \lambda_h, \nu_h, p_h, \dot{q}_h$ and $\dot{x}_h$ are Lipschitz continuous on $[0,T)$.

Corollary 3.4 implies that for any $z \in C(0,T)$ we can write

$$[\nu_h, z] = (\dot{\nu}_h, z) + \langle \nu_h(T)-\nu_h(T^-), z \rangle = (\dot{\nu}_h, z) - \langle \nu_h(T^-), z \rangle \ , \qquad (3.3.35)$$

$$\dot{\nu}_h(t) \geq 0 \quad \text{for almost all} \quad t \in [0,T], \qquad (3.3.35a)$$

$$-\nu_h(T^-) \geq 0. \qquad (3.3.35b)$$

From Corollary 3.6 and from (3.2.33) we obtain

Corollary 3.5

The Lagrange multiplier π_h in (3.2.27) is, after a possible correction on a set of measure zero, a Lipschitz continuous function an $[0,T]$.

3.4. Lipschitz Continuity with Respect to Parameters

In this section it will be shown that both the primal and dual optimal variables for Problems (OC_h) are Lipschitz continuous functions of the parameter h.

The presented results are due to A.L. Dontchev [12], also the proofs are only modifications of those given in [12].

Let us start with the following:

Lemma 3.7

For any compact set $\mathcal{H} \subset H$ there exists a constant $c > 0$ such that

$$||u_h||_\infty, ||\dot{x}_h||_\infty, ||\lambda_h||_\infty, ||\dot{\nu}_h||_1, |\nu_h(T^-)|, ||p_h||_\infty, ||\dot{q}_h||_\infty \leqslant c, \quad \forall h \in \mathcal{H}. \tag{3.4.1}$$

Proof

From (3.1.10), (3.2.17) and (3.3.35) for every $g \in S(h,\varepsilon(h))$ we obtain

$$F(u_g,x_g,g) = L(u_g,x_g;p_g,\lambda_g,\nu_g;g) \leqslant L(\hat{u}_h,x_g(\hat{u}_h);p_g,\lambda_g,\nu_g;g) =$$

$$= F(\hat{u}_h,x_g(\hat{u}_h),g) + (\lambda_g,\phi(\hat{u}_h,g)) + (\dot{\nu}_g,\theta(x_g(\hat{u}_h),g)) + \langle \nu_g(T^-),\theta(x_g(\hat{u}_h)(T),g)\rangle .$$

Taking into account (3.1.11), (3.2.20), (3.3.35a) and (3.3.35b) we get

$$0 \leqslant -\frac{\rho(h)}{2}\left[||\lambda_g||_1 + ||\dot{\nu}_g||_1 + |\nu_g(T^-)|\right] \leqslant F(\hat{u}_h,x_g(\hat{u}_h),g) - F(u_g,x_g,g) \quad \forall g \in S(h,\varepsilon(h)).$$

Hence by Lemma 3.1 we find that there exists a constant $c(h) > 0$ such that

$$||\lambda_g||_1, ||\dot{\nu}_g||_1, |\nu_g(T^-)| \leqslant c(h) \qquad \forall g \in S(h,\varepsilon(h)).$$

Since for any compact set $\mathcal{H}$ there exists an open convering consisting of a finite number of balls $S(h,\varepsilon(h))$ we get

$$||\lambda_h||_1, ||\dot{\nu}_h||_1, |\nu_h(T^-)| \leqslant c \qquad \forall h \in \mathcal{H}, \tag{3.4.2}$$

which in particular implies

$$||\nu_h||_\infty \leqslant c \qquad \forall h \in \mathcal{H}. \tag{3.4.3}$$

On the other hand by Lemma 3.1 and by the state equation (3.1.2) we get

$$||\dot{x}_h|| \leqslant c \qquad \forall h \in \mathcal{H} . \tag{3.4.4}$$

Applying Gronwall's lemma to (3.2.24) and using (3.4.3) and (3.4.2) we obtain

$$||q_h||_\infty \leqslant c \qquad \forall h \in \mathcal{H} , \tag{3.4.5}$$

which by (3.2.23) and (3.4.3) implies

$$||p_h||_\infty \leqslant c \qquad \forall h \in \mathcal{H} . \tag{3.4.6}$$

Now let us note that conditions of optimality (3.2.19), (3.2.20) implies that $u_h(t)$ and $\lambda_h(t)$ can be treated as the solution and the associated Lagrange multiplier of the following convex programming problem (QP_k):

$$\min_{u \in \Phi_h} \{f(u,x_h(t),h)-B^T(h)p_h(t)\},$$

where the vector $k^T \overset{def}{=} (x_h^T, p_h^T, h^T)$ plays the role of a parameter.

Note that by (3.4.4) and (3.4.6) we have

$$|k^T| \leqslant c \qquad \forall h \in \mathcal{H} .$$

On the other hand it is easy to see that for (QP_k) all assumptions of Lemma 1.1 are satisfied. Hence by Corollary 1.1 we obtain

$$|u_h(t)|, |\lambda_h(t)| \leqslant c \qquad \forall h \in \mathcal{H} ,$$

which implies

$$||u_h||_\infty, ||\lambda_h||_\infty \leqslant c \qquad \forall h \in \mathcal{H} . \tag{3.4.7}$$

Finally (3.1.2) and (3.2.24), together with (3.4.3) and (3.4.7) yield

$$||\dot{x}_h||_\infty, ||\dot{q}_h||_\infty \leqslant c \qquad \forall h \in \mathcal{H} . \qquad \square$$

Remark 3.1

From Lemma 3.7 and from the proof of Theorem 3.1 it follows that the Lipschitz modulus of the functions u_h, λ_h, ν_h and p_h can be chosen the same for all $h \in \mathcal{H}$.

To simplify notation, the subscripts 1 and 2 will be used in the sequal instead of h_1 and h_2.

The following lemma can be easily obtained from (3.1.2) using (C3) and Gronwall's lemma:

Lemma 3.8

For any compact set $\mathcal{H} \subset H$ there exists a constant $c > 0$ such that

$$||x_2 - x_1||_\infty \leq c\, [||u_2 - u_1|| + |h_2 - h_1|]. \tag{3.4.8}$$

Lemma 3.9

For any compact set $\mathcal{H} \in H$ there exists a constant $c > 0$ such that

$$||u_2 - u_1|| \leq c\, |h_2 - h_1|^{1/2} \quad \forall h_1, h_2 \in \mathcal{H}, \tag{3.4.9a}$$

$$||x_2 - x_1||_\infty \leq c\, |h_2 - h_1|^{1/2} \quad \forall h_1, h_2 \in \mathcal{H}. \tag{3.4.9b}$$

Proof

By (3.2.17) we have

$$L(u_2, x_2; p_1, \lambda_1, \nu_1; h_1) \geq L(u_1, x_1; p_2, \lambda_2, \nu_2; h_1). \tag{3.4.10}$$

Note that by (C1) and (C4) together with (3.2.20) and (3.2.21) the Lagrangian $L(\cdot,\cdot;p_2,\lambda_2,\nu_2,h_1)$ is a convex function. Expanding $L(\cdot,\cdot;$ $p_2,\lambda_2,\nu_2;h_1)$ into Taylor's series at (u_2,x_2) and using (3.1.5) we get

$$L(u_1,x_1;p_2,\lambda_2,\nu_2;h_1) \geq L(u_2,x_2;p_2,\lambda_2,\nu_2;h_1) + (D_u L(u_2,x_2;p_2,\lambda_2,\nu_2;h_1), u_1 - u_2) +$$
$$+ (D_x L(u_2,x_2;p_2,\lambda_2,\nu_2;h_1), x_1 - x_2) + \frac{\alpha}{2} \left\|u_1 - u_2\right\|^2. \tag{3.4.11}$$

Substituting (3.4.10) into (3.4.11) we obtain

$$||u_2 - u_1||^2 \leq c\{[L(u_2,x_2;p_1,\lambda_1,\nu_1;h_1) - L(u_2,x_2;p_2,\lambda_2,\nu_2;h_1)] +$$
$$+ (D_u L(u_2,x_2;p_2,\lambda_2,\nu_2;h_1), u_2 - u_1) + (D_x L(u_2,x_2;p_2,\lambda_2,\nu_2;h_1), x_2 - x_1)\}. \tag{3.4.12}$$

We are going to estimate all three terms on the right-hand side of (3.4.12).

Using definition (3.2.16) and (3.3.35) we get

$$L(u_2,x_2;p_1,\lambda_1,\nu_1;h_1) - L(u_2,x_2;p_2,\lambda_2,\nu_2;h_1) =$$
$$= (p_1 - p_2, \dot{x}_2 - A(h_1)x_2 - B(h_1)u_2) + (\lambda_1 - \lambda_2, \phi(u_2,h_1)) +$$
$$+ (\dot{\nu}_1 - \dot{\nu}_2, \theta(x_2,h_1)) - \langle \nu_1(T^-) - \nu_2(T^-), \theta(x_2(T),h_1)\rangle. \tag{3.4.13}$$

On the other hand

$$\dot{x}_2 - A(h_2)x_2 - B(h_2)u_2 = 0,$$

while by (3.2.20), (3.2.21) and (3.3.35)

$$(\lambda_1-\lambda_2,\phi(u_2,h_2)) \leqslant 0,$$

$$(\dot{\nu}_1-\dot{\nu}_2,\theta(x_2,h_2))-\nu_1(T^-)-\langle\nu_1(T^-),\theta(x_2(T),h_2)\rangle \leqslant 0 .$$

Hence

$$L(u_2,x_2;p_1,\lambda_1,\nu_1;h_1)-L(u_2,x_2;p_2,\lambda_2,\nu_2;h_1) \leqslant$$

$$\leqslant(p_1-p_2,(\dot{x}_2-A(h_1)x_2-B(h_1)u_2)-(\dot{x}_2-A(h_2)x_2 -B(h_2)u_2)) +$$

$$+(\lambda_1-\lambda_2,\phi(u_2,h_1)-\phi(u_2,h_2))+(\dot{\nu}_1-\dot{\nu}_2,\theta(x_2,h_1)-\theta(x_2,h_2)) +$$

$$-\langle\nu_1(T^-)-\nu_2(T^-),\theta(x_2(T),h_1)-\theta(x_2(T),h_2)\rangle . \qquad (3.4.14)$$

Taking into account (C3), (C5) and (3.4.1) we finally obtain

$$L(u_2,x_2;p_1,\lambda_1,\nu_1;h_1)-L(u_2,x_2;p_2,\lambda_2,\nu_2;h_1)\leqslant c|h_2-h_1| . \qquad (3.4.15)$$

Let us estimate two remaining terms in (3.4.12). Since (u_2,x_2) minimizes $L(.,.;p_2,\lambda_2,\nu_2;h_2)$ we have

$$(D_uL(u_2,x_2;p_2,\lambda_2,\nu_2;h_2),u_2-u_1) = 0, \qquad (3.4.16a)$$

$$(D_xL(u_2,x_2;p_2,\lambda_2,\nu_2;h_2),x_2-x_1) = 0. \qquad (3.4.16b)$$

Hence using (C2), (C3), (C5) as well as (3.4.1) we obtain from (3.2.16) and (3.4.16)

$$(D_uL(u_2,x_2;p_2,\lambda_2,\nu_2;h_1),u_2-u_1)=(D_uL(u_2,x_2;p_2,\lambda_2,\nu_2;h_1) +$$

$$-D_uL(u_2,x_2;p_2,\lambda_2,\nu_2;h_2),u_2-u_1) =$$

$$=(D_uF(u_2,x_2,h_2)-D_uF(u_2,x_2,h_1),u_2-u_1)+((B^T(h_1)-B^T(h_2))p_2,u_2-u_1) +$$

$$+((D_u\phi^T(u_2,h_2)-(D_u\phi^T(u_2,h_1))\lambda_2,u_2-u_1) \leqslant c|h_2-h_1|\,\|u_2-u_1\| . \qquad (3.4.17)$$

Similarly taking into account (3.3.35) and (3.4.8) we obtain

$$(D_xL(u_2,x_2;p_2,\lambda_2,\nu_2;h_1),x_2-x_1)=(D_xL(u_2,x_2;p_2,\lambda_2,\nu_2;h_1) +$$

$$-D_xL(u_2,x_2;p_2,\lambda_2,\nu_2;h_2),x_2-x_1) =$$

$$=(D_xF(u_2,x_2,h_2)-D_xF(u_2,x_2,h_1),x_2-x_1)+((A^T(h_1)-A^T(h_2))p_2,x_2-x_1) +$$

$$+((D_x\theta^T(x_2,h_2)-D_x\theta^T(x_2,h_1))\dot{\nu}_2,x_2-x_1) +$$

$$+ \langle (D_x\theta^T(x_2(T),h_2)-D_x\theta^T(x_2(T),h_1))\nu_2(T^-),x_2(T)-x_1(T)\rangle \leqslant$$

$$\leqslant c\ |h_2-h_1|\ \|x_2-x_1\|_\infty \leqslant c|h_2-h_1|(\ \|u_2-u_1\| + |h_2-h_1|). \qquad (3.4.18)$$

Substituting (3.4.15), (3.4.17) and (3.4.18) into (3.4.12) yields

$$\|u_2-u_1\| \leqslant c|h_2-h_1|^{1/2}.$$

The estimate (3.4.9b) follows from (3.4.8) and (3.4.9a).

□

Lemma 3.9 says that u_h is a Hölder continuous function of the parameter h with the exponent 1/2. We are going to show more, namely that it is a Lipschitz continuous function.

To this end we shall use again (3.4.12) but we have to apply more refine estimates of the terms on the right-hand side of this inequality.

The estimates (3.4.17) and (3.4.18) of the second and the third terms in (3.4.12) are good enough to obtain Lipschitz continuity, so we restrict ourself to the first term on the right-hand side od (3.4.12) Following [12] we get:

Lemma 3.10

For any $h_1,h_2\in\mathcal{H}$ we have

$$(\dot\nu_1-\dot\nu_2,\theta(x_2,h_1)-\theta(x_2,h_2))-\langle\nu_1(T)-\nu_2(T),\theta(x_2(T),h_1)-\theta(x_2(T),h_2)\rangle \leqslant$$

$$\leqslant c|h_2-h_1|(\ \|\nu_1-\nu_2\| +|h_2-h_1|). \qquad (3.4.19)$$

Proof.

By (C6) and (3.2.21) there exists $\tau>0$ such that

$$\nu_h(t) = \text{const} \qquad \forall t\in[0,\tau]\ , \qquad \forall h\in\mathcal{H}\ . \qquad (3.4.20)$$

Let us introduce a matrix function $W\in C^2(0,T;R^{s\times m})$ such that

$$W(0)=-D_h\theta(x_2(0),h_2)\ , \quad W(t)=0 \quad \text{for} \quad t\in[\tau,T].$$

Hence, integrating by parts we obtain

$$(\dot\nu_1-\dot\nu_2,\theta(x_2,h_1)-\theta(x_2,h_2))-\langle\nu_1(T)-\nu_2(T),\theta(x_2(T),h_1)-\theta(x_2(T),h_2)\rangle=$$

$$=(\dot\nu_1-\dot\nu_2,\theta(x_2,h_1)-\theta(x_2,h_2)+W(h_1-h_2))+$$

$$-\langle\nu_1(T)-\nu_2(T),\theta(x_2(T),h_1)-\theta(x_2(T),h_2)+W(T)(h_1-h_2)\rangle =$$

$$=-(\nu_1-\nu_2,[D_x\theta(x_2,h_1)-D_x\theta(x_2,h_2)]\dot x+\dot W(h_1-h_2)) +$$

$$+\langle\nu_1(0)-\nu_2(0),\theta(x_2(0),h_1)-\theta(x_2(0),h_2)+W(0)(h_1-h_2)\rangle \qquad (3.4.21)$$

Since by the definition of W it follows that

$$|\theta(x_2(0),h_1)-\theta(x_2(0),h_2)+W(0)(h_1-h_2)| \leqslant c|h_1-h_2|^2,$$

then, taking into account (3.4.1), we obtain (3.4.19) from (3.4.21). □

Substituting (3.4.19) into (3.4.14) and using (C3), (C5) as well as (3.4.1) we get

$$L_2(u_2,x_2;p_1,\lambda_1,\nu_1;h_1)-L(u_2,x_2;p_2,\lambda_2,\nu_2;h_1) <$$

$$\leqslant c|h_2-h_1|(\|p_1-p_2\| + \|\lambda_1-\lambda_2\| + \|\nu_1-\nu_2\| + |h_2-h_1|) \tag{3.4.22}$$

Comparing (3.4.12), (3.4.17), (3.4.18) and (3.4.22) it is easy to see that we would obtain Lipschitz continuity of u_h if we are able to estimate the distances of dual variables on the right-hand side of (3.4.22) in terms of $\|u_2-u_1\|$. To do that we shall need several lemmas.

For a given $h \in H$ and a fixed $\delta > 0$ let us introduce the set of the state constraints δ-binding at some $t \in [0,T]$, i.e.

$$J_h^\delta(t) = \{j \in J \mid \theta^j(x_h(t),h) \geqslant -\delta\} \tag{3.4.23a}$$

Moreover let us denote

$$K_h^\delta(t) = J \setminus J_h^\delta(t) = \{j \in J \mid \theta^j(x_h(t),h) < -\delta\} \tag{3.4.23b}$$

Lemma 3.11

There exists a constant $\delta > 0$, such that

$$|[D_u\phi^T_{I_h(t)}(u_h(t),h)-B^T(h)D_x\theta^T_{J_h^\delta(t)}(x_h(t),h)|v| \geqslant \frac{\beta}{2}|v| \tag{3.4.24}$$

for every $h \in \mathcal{H}$, every $t \in [0,T]$ and every v of appropriate dimension, where $D_x\theta^T_{J_h^\delta(t)}$ denotes the matrix whose columns are the gradients of all the functions θ^j, $j \in J_h^\delta(t)$.

Proof

Suppose that (3.4.24) is not true, then for any $\delta_h > 0$ there exists $h \in \mathcal{H}$, and $t_h \in [0,T]$ such that

$$\theta^j(x_h(t_h),h) \geqslant \delta_h \qquad \text{for} \qquad j \in J_h^{\delta_h}(t_h) \tag{3.4.25}$$

and for some v_h, $|v_h|=1$ we have

$$|[D_u\phi^T_{I_h(t_h)}(u_h(t_h),h),-B^T(h)D_x\theta^T_{J_h^{\delta_h}(t_h)}(x_h(t_h),h)]v_h| < \frac{\beta}{2}|v_h|. \qquad (3.4.26)$$

Let $\delta_h \to 0$.
From the sequence $\{h\}\subset \mathcal{H}$, we can extract a convergent subsequence, still denoted $\{h\}$, i.e. there exists $g\in \mathcal{H}$ such that

$$h \longrightarrow g.$$

The subsequence $\{h\}$ can be chosen in such way that

$$t_h \longrightarrow \bar{t}, \quad v_h \longrightarrow \bar{v} \quad \text{where} \quad |\bar{v}|=1,$$

and moreover the sets of indices $I_h(t_h)$ and $J_h^{\delta_h}(t_h)$ are independent of h:

$$I_h(t_h) = \bar{I} \ , \qquad J_h^{\delta_h}(t_h) = \bar{J} .$$

By Lemma 3.9 and Remark 3.1 we have

$$u_h(t_h) \longrightarrow u_g(\bar{t}) \quad \text{and} \quad x_h(t_h) \longrightarrow x_g(\bar{t})$$

Hence taking into account (3.4.25) we obtain

$$\phi^i(u_g(\bar{t}),g) = 0 \qquad \text{for} \qquad i\in \bar{I},$$

$$\theta^j(x_g(\bar{t}),g) = 0 \qquad \text{for} \qquad j\in \bar{J}.$$

i.e.

$$\bar{I}\subset I_g(\bar{t}), \qquad \bar{J}\subset J_g(\bar{t}).$$

On the other hand (3.4.26) together with (C3) and (C5) imply

$$|[D_u\phi^T_{\bar{I}}(u_g(\bar{t}),g),-B^T(g)D_x\theta^T_{\bar{J}}(x_g(\bar{t}),g)]\bar{v}| \leqslant \frac{\beta}{2}|\bar{v}| ,$$

which contradicts (C8). □

Let us define the sets

$$S^j = \{t\in [0,T]\,|\,\theta_1(x_1(t),h_1) \geqslant -\tfrac{\delta}{2}\} \qquad j\in J, \qquad (3.4.27)$$

where $\delta > 0$ is given by Lemma 3.11.
Let us put

$$\eta = \sup_{t,h,j} \{|D_x\theta^j(x_h(t),h)\dot{x}_h(t)|\}, \qquad (3.4.28)$$

where $t\in [0,T]$, $h\in \mathcal{H}$, $j\in J$. By (C5) and (3.4.1) η is finite.

Denote by $R^j_\gamma \subset S^j$ those subintervals for which

$$\text{meas } R^j_\gamma \geqslant \frac{1}{4}\frac{\delta}{\eta} \overset{\text{def}}{=} \gamma. \tag{3.4.29}$$

It is obvious that the number of subintervals R^j_γ is bounded uniformly for any $h \in \mathcal{H}$.

Let us define

$$R^j = \cup R^j_\gamma , \quad R = \bigcup_j R^j , \quad R_o = [0,T] \setminus R , \quad R_{\overline{J}} = \bigcap_{j \in \overline{J}} R^j , \tag{3.4.30}$$

where $\overline{J}$ is a subvector of J.

The interval $[0,T]$ can be split up into subintervals $R^k_{\overline{J}_k} = [t_{k-1}, t_k]$, $k=1,2,\dots,N$, including subintervals R^k_o. The number N of subintervals is bounded uniformly with respect to $h \in \mathcal{H}$.

By (3.4.27)-(3.4.30) it follows that

$$\theta^j(x_1(t),h_1) \leqslant -\frac{1}{4}\delta \qquad \forall t \in R_{\overline{J}_k} \quad \text{if} \quad j \notin \overline{J}_k. \tag{3.4.31a}$$

and moreover

$$\theta^j(x_1(t),h_1) \leqslant -\frac{1}{4}\delta \qquad \forall t \in [t_{k-1}, t_{k-1}+\gamma] \quad \text{if} \quad j \in \overline{J}_k \quad \text{and} \quad j \notin \overline{J}_{k-1}. \tag{3.4.31b}$$

By Lemma 3.9 and by (C5) there is a constant $\rho > 0$, such that

$$|\theta(x_2(t),h_2) - \theta(x_1(t),h_1)| \leqslant \frac{\delta}{8} \qquad \forall h_2 \in S(h_1,\rho), \quad \forall t \in [0,T]. \tag{3.4.32}$$

Hence if (3.4.32) holds, then by (3.4.27), (3.4.31) and (3.4.32) we have

$$\theta(x_2(t),h_2) \geqslant -\frac{5}{8}\delta \qquad \forall t \in R_{\overline{J}_k} \quad \text{if} \quad j \in \overline{J}_k , \tag{3.4.33a}$$

$$\theta(x_2(t),h_2) \leqslant -\frac{1}{8}\delta \qquad \forall t \in R_{\overline{J}_k} \quad \text{if} \quad j \notin \overline{J}_k , \tag{3.4.33b}$$

$$\theta(x_2(t),h_2) \leqslant -\frac{1}{8}\delta \qquad \forall t \in [t_{k-1}, t_{k-1}+\gamma] \quad \text{if} \quad j \in \overline{J}_k \quad \text{and} \quad j \notin \overline{J}_{k-1}. \tag{3.4.33c}$$

In the sequal by $\overline{J}(\cdot)$ and $\overline{K}(\cdot)$ there will be denoted the functions defined on $[0,T]$ which to every $t \in R^k_{\overline{J}_k}$ assign the corresponding multiindex $\overline{J}_k$ and the multiindex $\overline{K}_k = J \setminus \overline{J}_k$, respectively.

Vectors with subscripts $\overline{J}(t)$ or $\overline{K}(t)$ will denote the subvectors of s-dimensional vectors corresponding to the respective multiindices.

Lemma 3.12

If $h_2 \in S(h_1,\rho)$, then the following estimate holds

$$|\lambda_2(t)-\lambda_1(t)|,|\nu_{2,\bar{J}(t)}(t)-\nu_{2,\bar{J}(t)}(t)| \leqslant c[|h_2-h_1|+|u_2(t)-u_1(t)| + $$
$$+|x_2(t)-x_1(t)|+|q_2(t)-q_1(t)|+|\nu_{2,\bar{K}(t)}(t)-\nu_{2,\bar{K}(t)}(t)|] \quad \forall t\in [0,T]. \tag{3.4.34}$$

Proof

Let us rewrite the condition of optimality (3.2.25) and (3.2.10), at h_1 and h_2, in the form

$$D_u f(u_\ell(t), x_\ell(t),h_\ell)+B^T(h_\ell)[q_\ell(t)-D_x\theta^T_{\bar{K}(t)}(x_\ell(t),h_\ell)\nu_{\ell,\bar{K}(t)}(t)] +$$
$$-B^T(h_\ell)D_x\theta^T_{\bar{J}(t)}(x_\ell(t),h_\ell)\nu_{\ell,\bar{J}(t)}(t)+D_u\phi^T(u_\ell(t),h_\ell)\lambda_\ell(t) = 0 \tag{3.4.35a}$$

$$<\lambda_\ell(t),\ \phi(u_\ell(t),h_\ell)> = 0\ ,\quad \lambda_\ell(t) \geqslant 0 \qquad \ell=1,2. \tag{3.4.35b}$$

Moreover from (3.1.2) we obtain the following identity

$$D_x\theta_{\bar{J}(t)}(x_\ell(t),h_\ell)B(h_\ell)u_\ell(t)-D_x\theta_{\bar{J}(t)}(x_\ell(t),h_\ell)[\dot{x}_\ell(t)-A(h_\ell)x_\ell(t)] = 0$$
$$\ell=1,2 \tag{3.4.35b}$$

It is easy to see that by (3.4.35) u_ℓ and λ_ℓ, $\nu_{\ell,\bar{J}(t)}$, $\ell=1,2$, can be interpreted respectively as the unique solutions and the associated multipliers for the following parametric convex programming problem

$$\min\{f(u,x_\ell(t),h_\ell)+<B^T(h_\ell)[q_\ell(t)-D_x\theta^T_{\bar{K}(t)}(x_\ell(t),h_\ell)\nu_{\ell,\bar{K}(t)}(t)],u>\} \tag{3.4.36a}$$

subject to

$$\phi(u,h\ ) \leqslant 0, \tag{3.4.36b}$$

$$D_x\theta_{\bar{J}(t)}(x_\ell(t),h_\ell)B(h_\ell)u-D_xD_{\bar{J}(t)}(x_\ell(t),h_\ell)[\dot{x}_\ell(t)-A(h_\ell)x_\ell(t)]=0 \tag{3.4.36c}$$

Note that by Lemma 3.11 as well as by (3.4.27) and (3.4.33) we have

$$|[D_u\phi^T_{I_\ell(t)}(u_\ell(t),h_\ell),-B^T(h_\ell)D_x\theta^T_{\bar{J}(t)}(x_\ell(t),h_\ell)]v| \geqslant \frac{\beta}{2}|v| \quad \ell=1,2 \tag{3.4.37}$$

for every $t \in [0,T]$ and every v of appropriate dimension.

Taking into account (C1) through (C5) and (3.4.37) we find that all assumptions of Theorem 1.3 are satisfied for the problem (3.4.36) and by this theorem we have:

$$|u_2(t)-u_1(t)|,|\lambda_2(t)-\lambda_1(t)|,|\nu_{2,\bar{J}(t)}(t)-\nu_{1,\bar{J}(t)}(t)|\leqslant$$

$$\leqslant c\left[|x_2(t)-x_1(t)|+|q_2(t)-q_1(t)|+|\nu_{2,\bar{K}(t)}(t)-\nu_{1,\bar{K}(t)}(t)|+\right.$$

$$\left.+|(\dot{x}_2(t)-A(h_2)x_2(t))-(\dot{x}_1(t)-A(h_1)x_1(t))|+|h_2-h_1|\right].$$

Since $\dot{x}_\ell(t)-A(h_\ell)x_\ell(t)=B(h_\ell)u_\ell(t)$, $\ell=1,2$ we arrive at (3.4.34). □

In the next step we find the bounds for $|\nu_2(t)-\nu_1(t)|$.

Lemma 3.13

If $h_2\in S(h_1,\rho)$, then for any $t\in[t_{k-1},t_k]$, $k=1,2,\dots,N$ the following estimate holds:

$$|\nu_2(t)-\nu_1(t)|\leqslant c\sum_{j=k-1}^{N-1}(c+1)^{j-k+1}\left[|h_2-h_1|+\right.$$

$$+|u_2(t_j+(\tfrac{\gamma}{T})^{j-k+1}(t-t_{k-1}))-u_1(t_j+(\tfrac{\gamma}{T})^{j-k+1}(t-t_{k-1}))|+$$

$$+|x_2(t_j+(\tfrac{\gamma}{T})^{j-k+1}(t-t_{k-1}))-x_1(t_j+(\tfrac{\gamma}{T})^{j-k+1}(t-t_{k-1}))|+$$

$$\left.+|q_2(t_j+(\tfrac{\gamma}{T})^{j-k+1}(t-t_{k-1}))-q_1(t_j+(\tfrac{\gamma}{T})^{j-k+1}(t-t_{k-1}))|\right].\qquad(3.4.38)$$

Proof

We shall proceed by induction with respect to indices k of intervals $[t_{k-1},t_k]$ starting with $k=N$.

To simplify notation let us write

$$\nu=\nu_2-\nu_1,\quad \nu_{\bar{J}(t)}=\nu_{2,\bar{J}(t)}-\nu_{1,\bar{J}(t)},\quad \nu_{\bar{K}(t)}=\nu_{2,\bar{K}(t)}-\nu_{1,\bar{K}(t)},$$

$$z^T=[(u_2-u_1)^T,(x_2-x_1)^T,(q_2-q_1)^T],\quad h=h_2-h_1.\qquad(3.4.39)$$

By (3.2.21), (3.4.31a) and (3.4.33b) we have

$$\nu_{2,\bar{K}(t)}(t)=\nu_{1,\bar{K}(t)}(t)=\nu_{\bar{K}(t)}(t)=0\qquad\text{for}\quad t\in[t_{N-1},t_N].$$

Hence (3.4.34) implies

$$|\nu(t)|\leqslant c[|h|+|z(t)|]\qquad\forall t\in[t_{N-1},t_N].\qquad(3.4.40)$$

Now, let us proceed to the subinterval $[t_{N-2},t_{N-1}]$. If $J_N\subset J_{N-1}$, then the estimate (3.4.31b) holds on $[t_{N-2},t_{N-1}]$. Let us consider the case where $J_N\not\subset J_{N-1}$, hence by (3.4.31b) and (3.4.33c) we have

$$\nu_{\bar{K}(t)}(t)=\nu_{\bar{K}(t)}(s) \qquad \forall t\in [t_{N-2},t_{N-1}], \qquad \forall s\in [t_{N-1},t_{N-1}+\gamma].$$

Hence by (3.4.34) and (3.4.40) we obtain

$$|\nu(t)|=|\nu_{\bar{J}(t)}(t)|+|\nu_{\bar{K}(t)}(t)|\leqslant c[|h|+|z(t)|]+(c+1)|\nu_{\bar{K}(t)}(t)| =$$

$$=c[|h|+|z(t)|]+(c+1)|\nu_{\bar{K}(t)}(t_{N-1}+\tfrac{\gamma}{T}(t-t_{N-2})| \leqslant$$

$$\leqslant c[|h|+|z(t)|]+(c+1)|\nu(t_{N-1}+\tfrac{\gamma}{T}(t-t_{N-2}))| \leqslant$$

$$\leqslant c[1+(c+1)]|h|+c|z(t)|+(c+1)|z(t_{N-1}+\tfrac{\gamma}{T}(t-t_{N-2}))| \qquad \forall t\in [t_{N-2},t_{N-1}].$$

Similarly for $t\in [t_{N-3},t_{N-2}]$ we have

$$|\nu(t)| \leqslant c[|h|+|z(t)|]+(c+1)|\nu_{\bar{K}(t)}(t)| \leqslant$$

$$\leqslant c[|h|+|z(t)|]+(c+1)|\nu(t_{N-2}+\tfrac{\gamma}{T}(t-t_{N-3}))| \leqslant$$

$$\leqslant c[|h|+|z(t)|]+(c+1)[c+c(c+1)]|h|+c|z(t_{N-2}+\tfrac{\gamma}{T}(t-t_{N-3}))| +$$

$$+ (c+1)|z(t_{N-1}+(\tfrac{\gamma}{T})^2(t-t_{N-3}))| =$$

$$= c[1+(c+1)+(c+1)^2]|h|+c|z(t)|+c(c+1)|z(t_{N-2}+\tfrac{\gamma}{T}(t-t_{N-3}))| +$$

$$+ c(c+1)^2|z(t_{N-1}+(\tfrac{\gamma}{T})^2(t-t_{N-3}))|.$$

Proceding by induction we obtain

$$|\nu(t)| \leqslant c\sum_{j=k-1}^{N-1}(c+1)^{j-k+1}[|h|+|z(t_j+(\tfrac{\gamma}{T})^{j-k+1}(t-t_{k-1})|] \qquad \text{for } t\in [t_{k-1},t_k]$$

which by the definitions (3.4.39) gives (3.4.38). □

Now we are in a position to obtain the estimates of the dual variables:

<u>Lemma 3.14</u>

If $h_2\in S(h_1,\rho)$, then

$$||q_2-q_1||_\infty \leqslant c[|h_2-h_1|+||u_2-u_1||], \tag{3.4.41a}$$

$$||p_2-p_1|| \leqslant c[|h_2-h_1|+||u_2-u_1||], \tag{3.4.41b}$$

$$||\lambda_2-\lambda_1|| \leqslant c[|h_2-h_1|+||u_2-u_1||], \tag{3.4.41c}$$

$$||\nu_2-\nu_1|| \leqslant c[|h_2-h_1|+||u_2-u_1||]. \tag{3.4.41d}$$

Proof

Let us denote

$$q(t)=q_2(t)-q_1(t).$$

Subtracting the adjoint equations (3.2.24) at h_2 and h_1 respectively and integrating over $[t,T]$ we obtain

$$q(t)=\int_t^T\{-A^T(h_2)q(\tau)+[A^T(h_1)-A^T(h_2)]q_1(\tau)+C_2(\tau)[\nu_2(\tau)-\nu_1(\tau)]+$$
$$+[C_2(\tau)-C_1(\tau)]\nu_1(\tau)-[D_xf(u_2(\tau),x_2(\tau),h_2)-D_xf(u_1(\tau),x_1(\tau),h_1)]\}d\tau \tag{3.4.42}$$

where

$$C_\ell(t)=A^T(h_\ell)D_x\theta^T(x_\ell(t),h_\ell)+D_{xx}^2\theta^T(x_\ell(t),h_\ell)\dot{x}_\ell(t),\quad \ell=1,2.$$

Note that by (C2), (C3), (C5) and (3.1.2) as well as by Lemma 3.7 we have

$$|(A^T(h_1)-A^T(h_2))q_1(\tau)|\leqslant c|h_2-h_1|,$$

$$|[C_2(t)-C_1(t)]\nu_1(t)|,|D_xf(u_2(t),x_2(t),h_2)-D_xf(u_1(t),x_1(t),h_1)|\leqslant$$
$$\leqslant c[|h_2-h_1|+|u_2(t)-u_1(t)|+|x_2(t)-x_1(t)|],$$
$$|A^T(h_2)|,|C_2(t)|\leqslant c \qquad \text{for all} \qquad t\in[0,T]. \tag{3.4.43}$$

On the other hand, using the notation (3.4.39), from Lemma 3.13 we obtain for any $t\in[t_{k-1},t_k]$:

$$\int_t^T|\nu_2(t)-\nu_1(t)|dt=c\sum_{j=k-1}^{N-1}(c+1)^{j-k+1}[\int_t^T|h_2-h_1|d\tau+$$
$$+c\sum_{j=k-1}^{N-1}(c+1)^{j-k+1}\int_t^{t_k}|z(t_j+(\tfrac{\gamma}{T})^{j-k+1}(\tau-t_{k-1})|d\tau+$$
$$+c\sum_{m=k+1}^{N}\sum_{j=m-1}^{N-1}(c+1)^{j-m+1}\int_{t_{m-1}}^{t_m}|z(t_j+(\tfrac{\gamma}{T})^{j-m+1}(\tau-t_{m-1})|d\tau=$$

$$=c(T-t)\sum_{j=k-1}^{N-1}(c+1)^{j-k+1}|h_2-h_1|+c\sum_{j=k-1}^{N-1}(c+1)^{j-k+1}\int_{t_j+(\frac{\gamma}{T})^{j-k+1}(t-t_{k-1})}^{t_j+(\frac{\gamma}{T})^{j-k+1}(t_k-t_{k-1})}|z(\tau)|d\tau+$$
$$+\sum_{m=k+1}^{N}c\sum_{j=m-1}^{N-1}(c+1)^{j-m+1}\int_{t_j}^{t_j+(\frac{\gamma}{T})^{j-m+1}(t_m-t_{m-1})}|z(\tau)|d\tau. \tag{3.4.44}$$

Hence for any $t \in [0,T]$

$$\int_t^T |\nu_2(\tau)-\nu_1(\tau)|d\tau \leqslant c_1[T|h_2-h_1| + \int_t^T (|u_2(\tau)-u_1(\tau)|+|x_2(\tau)-x_1(\tau)| + $$
$$+ |q_2(\tau)-q_1(\tau)|)d\tau] , \qquad (3.4.45)$$

where $c_1 = c \sum_{j=0}^{N-1} (c+1)^j$.

Taking advantage of (3.4.42) and (3.4.45) we obtain from (3.4.42)

$$|q(t)| \leqslant c \int_t^T (|q(\tau)|+|h_2-h_1|+|u_2(\tau)-u_1(\tau)|+|x_2(\tau)-x_1(\tau)|)d\tau ,$$

which by Gronwall's lemma implies

$$|q_2(t)-q_1(t)| \leqslant c \int_t^T (|h_2-h_1|+|u_2(\tau)-u_1(\tau)|+|x_2(\tau)-x_1(\tau)|)d\tau .$$

Using (3.4.8) we obtain (3.4.41a).
By Lemma 3.13 and an evaluation similar to (3.4.44) we find that

$$\|\nu_2-\nu_1\| \leqslant c(|h_2-h_1|+ \|u_2-u_1\| + \|x_2-x_1\| + \|q_2-q_1\|),$$

which by (3.4.8) and (3.4.41a) implies (3.4.41d).
Finally (3.4.41a) and (3.4.41d) together with (3.2.23) and (3.4.34) imply (3.4.41b) and (3.4.41c) respectively. □

Since any compact set $\mathcal{H} \subset H$ can be covered by a finite number of balls $S(h,\rho)$ we obtain:

Corollary 3.6

For any compact and convex set $\mathcal{H} \subset H$ there exists a constant $c > 0$ such that

$$\|p_2-p_1\| , \|\lambda_2-\lambda_1\| , \|\nu_2-\nu_1\| \leqslant c [|h_2-h_1|+ \|u_2-u_1\|] \quad \forall h_1,h_2 \in \mathcal{H} \qquad (3.4.46)$$

Using (3.4.12), (3.4.17), (3.4.18), (3.4.22) and (3.4.46) we obtain the following principal result concerning Lipschitz continuity, with respect to the parameters, of the solutions and the Lagrange multipliers for (OC_h):

Theorem 3.2

If conditions (C1) through (C8) hold, then for any compact and convex set $\mathcal{H} \subset H$ there exists a constant $c > 0$ such that

$$\|u_2-u_1\|\,,\ \|\dot{x}_2-\dot{x}_1\|\,,\ \|p_2-p_1\|\,,\ \|\lambda_2-\lambda_1\|\,,\ \|\nu_2-\nu_1\| \leq c\,|h_2-h_1|$$

$$\text{for all}\quad h_1,h_2\in\mathcal{H}. \tag{3.4.47}$$

Note that by (3.4.20) for all $h\in\mathcal{H}$ we have

$$\nu_h(t) = \nu_h(0) \qquad \forall t\in(0,\tau).$$

Hence we obtain

$$|\nu_2(0)-\nu_1(0)| \leq \tau^{-1/2}\,\|\nu_2-\nu_1\| \qquad \forall h_1,h_2\in\mathcal{H}\,. \tag{3.4.48}$$

Conditions (3.2.33) together with (3.4.47) and (3.4.48) imply

<u>Corollary 3.7</u>

For any compact and convex set $\mathcal{H}\subset H$ there exists constants $c>0$ such that

$$\|\pi_2-\pi_1\|\,,\,|\sigma_2-\sigma_1| \leq c\,|h_2-h_1| \qquad \forall h_1,h_2\in\mathcal{H}. \tag{3.4.49}$$

Moreover Theorem 3.2 and Corollary 3.6 imply

<u>Corollary 3.8</u>

For any $h\in H$ the Lagrange multipliers p_h,λ_h,ν_h as well as π_h,σ_h are defined uniquely.

PART II

DIFFERENTIAL STABILITY

4. DIFFERENTIAL STABILITY OF SOLUTIONS TO CONVEX PROGRAMMING PROBLEMS

This chapter is devoted to studying properties of differentiability with respect to a parameter h of the solutions to the convex programming problems (P_h) defined in Chapter 1.

Most attention is devoted to proving that the solutions to (P_h) as well as the associated Lagrange multipliers are directionally differentiable functions. The form of right-differentials for these functions is derived.

Moreover conditions of continuous Gâteaux differentiability are obtained and the form of Clarke's generalized derivative is discussed.

4.1. Right-Differentiability of Solutions

Let us recall Family $\{P_n\}$ of Problems (P_h) depending on a vector parameter $h \in H \subset R^m$, which was defined in Chapter 1:

$$(P_h) \quad \left| \begin{array}{l} \text{find } \; u(h) \in R^n \; \text{ such that} \\ f(u(h),h) = \min\limits_{u \in \phi_h} f(u,h), \end{array} \right. \tag{4.1.1}$$

where

$$\Phi_h = \{u \in R^n \mid \phi^i(u,h) \leq 0, \; i \in I\}, \tag{4.1.2}$$

$$I = \{1,2,\ldots,r\}.$$

It is assumed that conditions (A1) through (A6) of Chapter 1 hold. Hence (P_h) has a unique solution and the associated Lagrange multipliers $\lambda(h)$ are also defined uniquely.

Like in Chapter 1 we denote

$$I_h = \{i \in I \mid \phi^i(u(h),h) = 0\}. \tag{4.1.3a}$$

Moreover we define the set $I_h^c \subset I_h$ of those binding constraints for which the strict complementarity condition holds:

$$I_h^c = \{i \in I_h \mid \lambda^i(h) > 0\}. \tag{4.1.3b}$$

We are going to show that the solutions $u(.)$ of (P_h) and the associated Lagrange multipliers $\lambda(.)$ are right-differentiable functions of h, i.e. for any $h \in H$ and any direction $g \in R^m$, $|g| = 1$, there exist the limits

$$\delta_h^+ u(h;g) = \lim_{\alpha \downarrow 0} \frac{1}{\alpha} [u(h+\alpha g) - u(h)] \tag{4.1.4a}$$

and

$$\delta_h^+\lambda(h;g) = \lim_{\alpha\downarrow 0} \frac{1}{\alpha}\left[\lambda(h+\alpha g) - \lambda(h)\right]. \tag{4.1.4b}$$

The following fundamental result is due to K. Jittorntrum [30]:

Theorem 4.1

If assumptions (A1) through (A6) hold, then the solutions $u(.)$ of (P_h) and the associated Lagrange multipliers $\lambda(.)$ are directionally differentiable functions of h at any $h \in H$, in any direction $g \in R^m$, $|g|=1$.

The right-differentials $v(h,g) \overset{\text{def}}{=} \delta_h^+u(h;g)$ and $\mu(h,g) \overset{\text{def}}{=} \delta_h^+\lambda(h;g)$ are given respectively by the solution and the associated Lagrange multipliers for the following quadratic programming problem:

$$(QP_{h,g}) \quad \left| \begin{array}{l} \text{find } v(h;g) \in R^n \text{ such that} \\ K(v(h,g),h,g) = \min\limits_{v \in \Psi_{h,g}} k(v,h,g), \end{array} \right. \tag{4.1.5}$$

where

$$k(v;h,g) = \frac{1}{2} <v,Q(h)v> + <q(h,g),v>, \tag{4.1.6}$$

$$Q(h) = D^2_{uu}L(u(h),\lambda(h),h) = D^2_{uu}f(u(h),h) + \sum_{i\in I} \lambda^i(h)D^2_{uu}\phi^i(u(h),h), \tag{4.1.6a}$$

$$q(h,g) = D^2_{uh}L(u(h),\lambda(h),h)g =$$

$$= \left[D^2_{uh}f(u(h),h) + \sum_{i\in I}\lambda^i(h)D^2_{uh}\phi^i(u(h),h)\right]g, \tag{4.1.6b}$$

$$\Psi_{h,g} = \{v \in R^n \mid \psi^i(v,h,g) \overset{\text{def}}{=} <D_u\phi^i(u(h),h),v> + <D_h\phi^i(u(h),h),g> \begin{cases} =0 & \text{for } i \in I_h^c \\ \leq 0 & \text{for } i \in I_h \setminus I_h^c \end{cases}\}. \tag{4.1.7}$$

Moreover

$$\mu^i(h;g) \overset{\text{def}}{=} \delta_h^+\lambda^i(h;g) = 0 \quad \text{for} \quad i \in I\setminus I_h. \tag{4.1.8}$$

Lagrangian $L(u,\lambda;h)$ is defined here by (1.1.12).

Proof

By (A.1), (A.3) and (1.1.16) the matrix $Q(h)$ is positive definite, while by (A.6) the set $\Psi_{h,g}$ is non-empty. Hence the solution of $(QP_{h,g})$ exists and is unique. By (A.6) the associated Lagrange multipliers are unique.

The solution to $(QP_{h,g})$ is characterized by the Kuhn-Tucker conditions:

$$Q(h)v(h,g)+q(h,g)+\sum_{i\in I}\mu^i(h,g)D_u\phi^i(u(h),h) = 0, \tag{4.1.9a}$$

$$\mu^i(h,g)\left[\langle D_u\phi^i(u(h),h),v(h,g)\rangle+\langle D_h\phi^i(u(h),h),g\rangle\right]=0 \quad i\in I_h\setminus I_h^c, \tag{4.1.9b}$$

$$\mu^i(h,g) \geq 0 \qquad i\in I_h\setminus I_h^c, \tag{4.1.9c}$$

where

$$\langle D_u\phi^i(u(h),h),v(h,g)\rangle + \langle D_h\phi^i(u(h),h),g\rangle=0 \quad i\in I_h^c, \tag{4.1.10a}$$

$$\langle D_u\phi^i(u(h),h),v(h,g)\rangle + \langle D_h\phi^i(u(h),h),g\rangle \leq 0 \quad i\in I_h\setminus I_h^c, \tag{4.1.10b}$$

$$\mu^i(h,g) = 0 \qquad i\in I\setminus I_h. \tag{4.1.10c}$$

Let $\{\alpha\}\downarrow 0$ be any arbitrary sequence. By Theorem 1.3 we have

$$\left|\frac{1}{\alpha}\left[u(h+\alpha g) - u(h)\right]\right| \leq c,$$

$$\left|\frac{1}{\alpha}\left[\lambda(h+\alpha g) - \lambda(h)\right]\right| \leq c.$$

Hence we can extract a subsequence $\{\alpha'\}\subset\{\alpha\}$, such that

$$\frac{1}{\alpha'}\left[u(h+\alpha' g) - u(h)\right] \to v, \tag{4.1.11a}$$

$$\frac{1}{\alpha'}\left[\lambda(h+\alpha' g) - \lambda(h)\right] \to \mu. \tag{4.1.11b}$$

We are going to prove that $v\in R^n$ and $\mu\in R^r$ given by (4.1.11) satisfy (4.1.9) and (4.1.10), hence they are defined uniquely and are independent of the choice of sequences $\{\alpha\}$ and $\{\alpha'\}$. Therefore v and μ are respective right-differentials.

To show that v and μ satisfy (4.1.9) we use Kuhn-Tucker conditions (1.1.14)-(1.1.16) for (P_h). Taking the difference quotient of (1.1.14) at $(h+\alpha' g)$ and at h, passing to the limit with $\alpha'\downarrow 0$ and using (A1) through (A4), as well as (4.1.6) and (4.1.10) we arrive at (4.1.9a).

Now let us note that by (4.1.3a)

$$\phi^i(u(h),h) = 0 \quad \text{for} \quad i\in I_h, \tag{4.1.12}$$

while

$$\phi^i(u(h+\alpha' g),h+\alpha' g) \leq 0 \quad \text{for} \quad i\in I.$$

Taking the difference quotient of these formulas, using (A3), (A4), (4.1.11a) and letting $\alpha'\downarrow 0$ we find that (4.1.10b) holds for $i\in I_h$.

To show (4.1.10a) note that by (4.1.3b)

$$\lambda^i(h) > 0 \quad \text{for} \quad i \in I_h^c.$$

Hence by Theorem 1.3, for $\alpha' > 0$ sufficiently small, we get

$$\lambda^i(h+\alpha' g) > 0 \quad \text{for} \quad i \in I_h^c$$

and by (1.1.15)

$$\phi^i(u(h+\alpha' g), h+\alpha' g) = 0 \quad \text{for} \quad i \in I_h^c. \tag{4.1.13}$$

Equalities (4.1.12) and (4.1.13) yield (4.1.10a).

Now we shall show (4.1.9c). By (4.1.3) we have

$$\lambda^i(h) = 0 \quad \text{for} \quad i \in I_h \setminus I_h^c.$$

On the other hand

$$\lambda^i(h+\alpha' g) \geq 0 \quad \text{for} \quad i \in I.$$

These two formulas, together with (4.1.11b) imply (4.1.9c). To prove (4.1.9b) it is enough to show that for $i \in I_h \setminus I_h^c$

$$< D_u\phi^i(u(h),h), v > + < D_h\phi^i(u(h),h), g > < 0 \quad \text{implies} \quad \mu^i = 0. \tag{4.1.14}$$

By (4.1.3)

$$\phi^i(u(h),h) = 0, \tag{4.1.15a}$$

$$\lambda^i(h) = 0 \qquad \text{for} \quad i \in I_h \setminus I_h^c. \tag{4.1.15b}$$

If $< D_u\phi^i(u(h),h), v> + < D_h\phi^i(u(h),h), g >< 0$ then by (A3), (A4), (4.1.11a) and (4.1.15a) we get

$$\phi^i(u(h+\alpha' g), h+\alpha' g) < 0 \quad \text{for} \quad \alpha' > 0 \quad \text{sufficiently small.}$$

By (1.1.15) this implies

$$\lambda^i(h+\alpha' g) = 0 \quad \text{for} \quad \alpha' > 0 \quad \text{sufficiently small,} \tag{4.1.16}$$

which together with (4.1.15b) and (4.1.11b) yield (4.1.14).
To complete the proof of the theorem it remains to show that (4.1.10c) holds.

From (4.1.3a) and (1.1.15) it follows that for $i \in I \setminus I_h$

$$\phi^i(u(h),h) < 0, \tag{4.1.17a}$$

$$\lambda^i(h) = 0. \tag{4.1.17b}$$

Using Theorem 1.3 as well as (A3),(A4) and (4.1.17a) we get

$$\phi^i(u(h+\alpha' g),h+\alpha' g) < 0 \quad \text{for} \quad \alpha' > 0 \quad \text{sufficiently small,}$$

which by (1.1.15) implies

$$\lambda^i(h+\alpha' g) = 0. \tag{4.1.18}$$

Equalities (4.1.17b) and (4.1.18) yield (4.1.10c). □

Note that condition of optimality (4.1.9a) for $(QP_{h,g})$ can be expressed in term of Lagrangian $L(u,\lambda;h)$ in the following simple form

$$D^2_{uu}L(u(h),\lambda(h);h)v(h,g) + D^2_{u\lambda}L(u(h),\lambda(h),h)\mu(h,g) +$$
$$+ D^2_{uh}L(u(h),\lambda(h);h)g = 0. \tag{4.1.19}$$

Remark 4.1

As in the case of Theorem 1.3 also Theorem 4.1 remains true if Φ_h contains, along with inequality type constraints, also affine equality type constraints, provided that (A6) is satisfied for all binding constraints. In $(QP_{h,g})$ all equality type constraints belong to the set I^c_h.

4.2. Estimates of the Rate of Convergence

In this section it will be shown that, under stronger assumptions on data, the rate of convergence in (4.1.11) can be estimated.

Instead of (A1) through (A4) we introduce the following conditions, where $k \geq 1$ is a fixed integer:

(A1') for each $h \in H$ $f(\cdot,h)$ is $(k+1)$- times continuously differentiable function of u. Moreover it is strongly convex, uniformly with respect to h, i.e. there exists a constant $\alpha > 0$ independent of h, such that

$$< v, D^2_{uu}f(u,h)v > \geq \alpha|v|^2 \qquad \forall u,v \in R^n, \quad \forall h \in H,$$

(A2') $f(\cdot,\cdot)$ and $D_u f(\cdot,\cdot)$ are k-times continuously differentiable functions on $R^n \times H$,

(A3') for each $h \in H$ $\phi^i(\cdot,h)$, $i \in I$ are $(k+1)$- times continuously differentiable and convex functions of u,

A4' $\phi^i(\cdot,\cdot)$ and $D_u\phi^i(\cdot,\cdot)$, $i \in I$, are k-times continuously differentiable functions on $R^n \times H$

Note that for $k=1$ conditions (A1')-(A4') coincide with (A1)-(A4).

Let us introduce the sets

$$J_{h,g} = \{i \in I_h \mid \psi^i(v(h,g);h,g) = 0\}, \tag{4.2.1}$$

$$K_{h,g} = I_h^c \cup \{i \in I_h \setminus I_h^c \mid \mu^i(h,g) > 0\}. \tag{4.2.2}$$

We shall need the following auxiliary

Lemma 4.1

For any $h \in H$ and any $g \in R^m$, $|g|=1$ there exists $\alpha_o > 0$ such that for all $\alpha \in (0,\alpha_o)$ the following inclusions take place

$$I_h^c \subset K_{h,g} \subset I_{h+\alpha g}^c \subset I_{h+\alpha g} \subset J_{h,g} \subset I_h. \tag{4.2.3}$$

Proof

By definitions (4.1.3) and by Theorem 1.3, for $\alpha > 0$ sufficiently small we have

$$I_h^c \subset I_{h+\alpha g}^c \subset I_{h+\alpha g} \subset I_h.$$

Hence to prove (4.2.3) it is enough to show that

$$K_{h,g} \subset I_{h+\alpha g}^c \quad \text{and} \quad I_{h+\alpha g} \subset J_{h,g}. \tag{4.2.4}$$

Let $i \in I_h \setminus J_{h,g}$, i.e.

$$\phi^i(u(h),h) = 0$$

and

$$< D_u\phi^i(u(h),h),v(h) > + <D_h\phi^i(u(h),h),g> < 0,$$

then by (4.1.4a) and by (A4), for $\alpha > 0$ sufficiently small, we have

$$\phi^i(u(h+\alpha g),h+\alpha g) < 0,$$

i.e. $i \notin I_{h+\alpha g}$, which implies the right inclusion in (4.2.4).

Now let $i \in I_h \setminus I_h^c$ and $\mu^i(h,g) > 0$, then by (4.1.3) and (4.1.4b), for $\alpha > 0$ sufficiently small, we have

$$\lambda^i(h+\alpha g) > 0,$$

i.e. $i \in I_{h+\alpha g}^c$, which implies the left inclusion in (4.2.4). Hence we can find $\alpha_o > 0$, such that (4.2.3) holds.

□

Let us denote

$$\eta(\alpha) = \frac{1}{\alpha}\left[u(h+\alpha g) - u(h)\right] - v(h,g), \tag{4.2.5a}$$

$$\kappa(\alpha) = \frac{1}{\alpha}\left[\lambda(h+\alpha g) - \lambda(h)\right] - \mu(h,g), \tag{4.2.5b}$$

Taking advantage of (4.2.3) we find that for $\alpha > 0$ sufficiently small we have

$$\kappa^i(\alpha) = 0 \quad \text{for} \quad i \notin I_{h+\alpha g} \tag{4.2.6}$$

and

$$\frac{1}{\alpha}\left[\phi^i(u(h+\alpha g),h+\alpha g)-\phi^i(u(h),h)\right]-\left[< D_u\phi^i(u(h),h)v(h,g) > + \right.$$

$$\left. + < D_h\phi^i(u(h),h),g >\right] = 0 \quad \text{for} \quad i \in I_{h+\alpha g}. \tag{4.2.7}$$

Expanding $\phi^i(.,.)$ into Taylor's series at the point $(u(h),h)$ we get

$$\phi^i(u(h+\alpha g),h+\alpha g)-\phi^i(u(h),h) =$$

$$= \alpha < D_u\phi^i(u(h),h),\frac{1}{\alpha}(u(h+\alpha g)-u(h) > + \alpha < D_h\phi^i(u(h),h),g > -\alpha^2 b^i, \tag{4.2.8}$$

where

$$b^i = -\frac{1}{2}\left[\frac{1}{\alpha}(u^T(h+\alpha g)-u^T(h),g^T)\right]\left[\int_0^1 D^2\phi(u_t,h_t)(1-t)dt\right]\begin{bmatrix}\frac{1}{\alpha}(u(h+\alpha g)-u(h)) \\ \\ g\end{bmatrix}. \tag{4.2.9}$$

Here $u_t=tu(h+\alpha g)+(1-t)u(h)$, $h_t=t(h+\alpha g)+(1-t)h$, $t\in[0,1]$ and $D^2\phi(u,h)$ denotes the Hessian matrix of $\phi(u,h)$.

Substituting (4.2.8) into (4.2.7) and using (4.2.5a) we obtain

$$<D_u\phi^i(u(h),h),\eta(\alpha)> = \alpha b^i \quad \text{for} \quad i \in I_{h+\alpha g}. \tag{4.2.10}$$

Let us denote by D_uL^j - the j-th component of the n-dimensional vector D_uL.

Expanding $D_uL^j(.,.,.)$ into Taylor's series at $(u(h),\lambda(h),h)$ and using (1.1.14) we get

$$0=D_uL^j(u(h+\alpha g),\lambda(h+\alpha g),h+\alpha g)-D_uL^j(u(h),\lambda(h),h) =$$

$$= \alpha < D^2_{uu}L^j(u(h),\lambda(h),h),\frac{1}{\alpha}(u(h+\alpha g)-u(h))> + \tag{4.2.11 }$$

$$+ \alpha < D^2_{u\lambda}L^j(u(h),\lambda(h),h),\frac{1}{\alpha}(\lambda(h+\alpha g)-\lambda(h)) + \alpha < D^2_{uh}L^j(u(h),\lambda(h),h),g> - \alpha^2 a^j$$

where

$$a^j = -\frac{1}{2}\left[\frac{1}{\alpha}(u^T(h+\alpha g)-u^T(h)),\frac{1}{\alpha}(\lambda^T(h+\alpha g)-\lambda^T(h)),g^T\right] \times$$

$$\times \left[\int_0^1 D^2(D_u L^j(u_t,\lambda_t,h_t)(1-t)dt\right] \begin{bmatrix} \frac{1}{\alpha}(u(h+\alpha g)-u(h)) \\ \frac{1}{\alpha}(\lambda(h+\alpha g)-\lambda(h)) \\ g \end{bmatrix} \qquad (4.2.12)$$

Here u_t,λ_t and h_t are defined in the same way as in (4.2.9) and $D^2(D_uL^j(u,\lambda,h))$ denotes the Hessian matrix of $D_uL^j(u,\lambda,h)$. From (4.1.19) we have

$$\langle D^2_{uu}L^j(u(h),\lambda(h);h),\nu(h,g)\rangle+\langle D^2_{u\lambda}L^j(u(h),\lambda(h);h)\mu(h,g)\rangle+$$

$$+\langle D^2_{uh}L^j(u(h),\lambda(h);h),g\rangle = 0. \qquad (4.2.13)$$

Dividing (4.2.11) by α, subtracting from it (4.2.13) and using (4.2.5) we obtain

$$\langle D^2_{uu}L^j(u(h),\lambda(h);h),\eta(\alpha)\rangle+\langle D^2_{u\lambda}L^j(u(h),\lambda(h);h),\kappa(\alpha)\rangle=\alpha a^j,$$

$$j=1,2,\dots,n\,. \qquad (4.2.14)$$

Let us denote by $\tilde{\kappa}(\alpha)$ and $\tilde{\phi}(u,h)$ the subvectors of $\kappa(\alpha)$ and $\phi(u,h)$ respectively containing all components with indices $i\in I_{h+\alpha g}$. Using this notation and taking advantage of (4.2.6) we can rewrite (4.2.10) and (4.2.14) in the form of the following matrix equation

$$\begin{bmatrix} Q(h) & , D_u\tilde{\phi}^T(u(h),h) \\ D_u\tilde{\phi}(u(h),h), & 0 \end{bmatrix}\begin{bmatrix}\eta(\alpha)\\ \tilde{\kappa}(\alpha)\end{bmatrix} = \alpha\begin{bmatrix}a\\ b\end{bmatrix} \qquad (4.2.15)$$

where $Q(h)$ is given by (4.1.6), while a and b are the vectors of components a^j and b^i given by (4.2.12) and (4.2.9).

Note that the matrix on the left-hand side of (4.2.15) has the structure (1.4.8), where conditions (1.4.9) and (1.4.10) are satisfied by (A1), (A3), (A6) and (1.1.16). Hence by Lemma 1.2 and Corollary 1.1 we get

$$\left\| \begin{bmatrix} Q(h) & , D_u\tilde{\phi}^T(u(h),h) \\ D_u\tilde{\phi}(u(h),h), & 0 \end{bmatrix}^{-1} \right\| \leqslant c\,,$$

where c does not depend on g,α and $h\in\mathcal{H}$.

Let us assume that conditions (A1')-(A4') are satisfied with $k=2$. Using these conditions as well as Corollary 1.1 and Theorem 1.3 we get from (4.2.9) and (4.2.12)

$$|a|,\ |b| \leq c,$$

where c does not depend on g, α and $h \in \mathcal{H}$.
Hence from (4.2.15) we obtain

$$|\eta(\alpha)|,\ |\tilde{\kappa}(\alpha)| \leq c\,\alpha. \tag{4.2.16}$$

Thus, taking into account (4.2.6) and definitions (4.2.5) we arrive at

Theorem 4.2

If conditions (A1')-(A4'), with $k=2$, as well as (A5) and (A6) are satisfied, then for any $h \in \mathcal{H}$ and any direction $g \in R^m, |g|=1$, there exists a constant $\alpha_{h,g} > 0$ such that for all $\alpha \in (0, \alpha_{h,g})$ the following estimates hold

$$\left| \frac{1}{\alpha} \left[u(h+\alpha g)-u(h)\right]-v(h,g) \right| \leq c\,\alpha, \tag{4.2.17a}$$

$$\left| \frac{1}{\alpha} \left[\lambda(h+\alpha g)-\lambda(h)\right]-\mu(h,g) \right| \leq c\,\alpha, \tag{4.2.17b}$$

where c does not depend on g, α and $h \in \mathcal{H}$.

4.3. Higher Order Right-Differentiability of Solutions

In this section the problem of higher order right-differentiability of solutions to (P_h) and of the associated multipliers is considered.

We restrict our analysis to the second derivative, which requires assumptions (A1')-(A4') with $k=2$. However using exactly the same argument the form of higher order derivatives can be found.

Let us start with proving that

$$v(h,g) = \delta_h^+ u(h,g) \quad \text{and} \quad \mu(h,g) = \delta_h^+ \lambda(h,g),$$

as the functions of h, are Lipschitz continuous along the direction g. Namely the following result holds:

Proposition 4.1

If conditions (A1')-(A4'), with $k=2$, as well as (A5) and (A6) hold, then for every compact and convex set $\mathcal{H} \subset H$ there exists a constant $c > 0$ such that for every h, $h+\alpha g \in \mathcal{H}$, with $\alpha > 0$, we have

$$|v(h+\alpha g, g)-v(h,g)| \leq c\alpha \tag{4.3.1a}$$

$$|\mu(h+\alpha g,g)-\mu(h,g)| \leqslant c\alpha \qquad (4.3.1b)$$

Proof

By Theorem 4.1 $v(h,g)$ are given by the solutions to the quadratic programming problems $(QP_{h,g})$. To prove (4.3.1) we have to analyse the dependence of solutions to $(QP_{h,g})$ and of the associated multipliers on the parameter h. Note that we can not apply directly Theorem 1.3 since in the case of Problems $(QP_{h,g})$ also the sets I_h of constraint functions $\phi^i(\cdot,h)$ depend on h. However we shall use again the abstract Theorem 1.2.

By that theorem to prove (4.3.1) it is enough to show that:

1) $v(h+\alpha g,g)$ and $\mu(h+\alpha g,g)$ are continuous functions of α,

2) they are Lipschitz continuous, with the same Lipschitz modulus for all pairs $(h,h+\alpha g)\in \mathcal{H}\times\mathcal{H}$, such that $\{i\in I_h | \psi^i(v(h;g)h,g) = 0\} = \{i\in I_{h+\alpha g} | \psi^i(v(h+\alpha g;g),h+\alpha g,g) = 0\}$ and $\alpha>0$ is sufficiently small.

We shall prove that $v(h+\alpha g,g)$ and $\mu(h+\alpha g,g)$ are locally Lipschitz continuous functions of α, which implies 1).

Let us recall definitions (4.2.1) and (4.2.2) of the sets $J_{h,g}$ and $K_{h,g}$.

Note that for $i\in J_{h,g}\setminus K_{h,g}$ the constraints $\psi^i(v;h,g)$ do not influence the solution to $(QP_{h,g})$. Hence $v(h,g)$ can be treated as the solution to the problem $(QP^K_{h,g})$ formulated like $(QP_{h,g})$ but with the set $\Psi_{h,g}$ substituted by

$$\Psi^K_{h,g} = \{v\in R^n \mid \Psi^i(v;h,g) = 0 \quad \text{for} \quad i\in K\}, \qquad (4.3.2)$$

where K is any set such that

$$K_{h,g}\subset K\subset J_{h,g}. \qquad (4.3.3)$$

By (A6) for any set K satisfying (4.3.3) the gradients of the functions $\psi^i(.;h,g)$ are linearly independent for $i\in K$. Hence, by Theorem 1.3 and Remark 1.1 for fixed K the solutions to (QP^K_h) and the associated Lagrange multipliers are Lipschitz continuous on any compact subset of H. Of course we can choose a Lipschitz modulus independent of K satisfying (4.3.3).

On the other hand by (4.2.3) the set $K=I_{h+\alpha g}$ satisfy (4.3.3) for $\alpha > 0$ and sufficiently small. It shows that for such an α (4.3.1) holds.

To complete the proof of the proposition it remains to show 2).

It is done in exactly the same way as in the proof of Theorem 1.3. □

We are going to prove the existence and to find the form of

$$w(h,g) \overset{\text{def}}{=} \delta_h^+ v(h,g) = \delta_{hh}^{++} u(h,g,g) \tag{4.3.4a}$$

and

$$\nu(h,g) \overset{\text{def}}{=} \delta_h^+ \mu(h,g) = \delta_{hh}^{++} \lambda(h,g,g). \tag{4.3.4b}$$

To this purpose we shall need the following:

Lemma 4.2

Let $\xi(.)$ be a real-valued Lipschitz continuous function. If there exist two sequences $\{\alpha_i\}\downarrow 0$, $i=1,2$, such that

$$\lim_{\alpha_i \downarrow 0} \{ \frac{1}{\alpha_i} [\xi(a+\alpha_i)-\xi(a)]\} = \eta_i \ , \quad i=1,2$$

and $\eta_1 \neq \eta_2$,

then for any $\eta=\ell\eta_1+(1-\ell)\eta_2$, with $\ell \in (0,1)$ there exists a sequence $\{\alpha\}\downarrow 0$, such that

$$\lim_{\alpha\downarrow 0} \{ \frac{1}{\alpha} [\xi(a+\alpha)-\xi(a)]\} = \eta.$$

The proof of Lemma 4.2 follows immediately from the continuity of $\xi(.)$.

Let us define the following sets

$$N_{h,g}=\{i \in J_{h,g}\setminus K_{h,g} \mid \exists\, \alpha^i > 0 \ \text{ s.t. } \ \mu^i(h+\alpha g,g) < 0 \quad \forall \alpha \in (0,\alpha^i)\}, \tag{4.3.5a}$$

$$M_{h,g} = K_{h,g} \cup N_{h,g}, \tag{4.3.5b}$$

$$S_{h,g}=\{i \in J_{h,g}\setminus K_{h,g} \mid \exists\, \alpha^i > 0 \ \text{ s.t. } \ \psi^i(v(h+\alpha g),g) > 0 \quad \forall \alpha \in (0,\alpha^i)\}, \tag{4.3.6a}$$

$$R_{h,g} = J_{h,g}\setminus S_{h,g}. \tag{4.3.6b}$$

Theorem 4.3

If conditions (A1')-(A4'), with $k=2$, as well as (A5) and (A6) hold, then there exist the second right-differentials of $u(.)$ and $\lambda(.)$ at h in the direction g, defined by (4.3.4).

They are given respectively by the unique solution and the unique associated Lagrange multiplier of the following quadratic programming problem:

$$(QP^2_{h,g}) \quad \left| \begin{array}{l} \text{find } w(h,g) \in R^n \text{ such that} \\ k^{(2)}(w(h,g),h,g) = \min\limits_{w \in X_{h,g}} k^{(2)}(w,h,g), \end{array} \right. \tag{4.3.7}$$

where

$$k^{(2)}(w,h,g) = \frac{1}{2} < w, Q(h)w > + < r(h,g), w >, \tag{4.3.8}$$

$$Q(h) = D^2_{uu} f(u(h),h) + \sum_{i \in I} \lambda^i(h) D^2_{uu} \phi^i(u(h),h), \tag{4.3.8a}$$

$$r(h,g) = [D_h Q(h)g] v(h,g) + D_h q(h,g) g +$$

$$+ \sum_{i \in I} \mu^i(h,g) [D^2_{uu} \phi^i(u(h),h) v(h,g) + D^2_{uh} \phi^i(u(h),h) g] =$$

$$= [D^3_{uuu} f(u(h),h) v(h,g)] v(h,g) + 2 [D^3_{uuh} f(h(h),h) v(h,g)] g + [D^3_{uhh} f(u(h),h) g] g +$$

$$+ \sum_{i \in I} \{\lambda^i(h) ([D^3_{uuu} \phi^i(u(h),h) v(h,g)] v(h,g) + 2 [D^3_{uuh} \phi^i(u(h),h) v(h,g)] g +$$

$$+ [D^3_{uhh} \phi^i(u(h),h) g] g) + 2\mu^i(h,g) (D^2_{uu} \phi^i(u(h),h) v(h,g) + D^2_{uh} \phi^i(u(h),h) g)\}, \tag{4.3.8b}$$

$$X_{h,g} = \{ w \in R^n \mid \chi^i(w,h,g) \begin{cases} = 0 & \text{for } i \in M_{h,g} \\ \leq 0 & \text{for } i \in R_{h,g} \setminus M_{h,g} \end{cases} \} \tag{4.3.9}$$

$$\chi^i(w,h,g) \overset{\text{def}}{=} < D_u \phi^i(u(h),h), w > +$$

$$+ [< D^2_{uu} \phi^i(u(h),h) v(h,g) + D^2_{uh} \phi^i(u(h),h) g, v(h,g) > +$$

$$+ < D^2_{uh} \phi^i(u(h),h) v(h,g) + D^2_{hh} \phi^i(u(h),h) g, g >]. \tag{4.3.9a}$$

Moreover

$$\nu^i(h,g) = 0 \quad \text{for} \quad i \notin R_{h,g}. \tag{4.3.10}$$

<u>Proof</u>

The proof of the theorem will be very similar to that of Theorem 4.1. The Kuhn-Tucker conditions for $(QP^{(2)}_{h,g})$ can be expressed in the form

$$Q(h)w(h,g) + r(h,g) + \sum_{i \in I} \nu^i(h,g) D_u \phi^i(u(h),h) = 0, \tag{4.3.11a}$$

$$\nu^i(h,g) \chi^i(w(h,g),h,g) = 0 \qquad i \in R_{h,g} \setminus M_{h,g}, \tag{4.3.11b}$$

$$\nu^i(h,g) \geq 0 \qquad i \in R_{h,g} \setminus M_{h,g}, \tag{4.3.11c}$$

where

$$\chi^{i}(w(h,g),h,g) = 0 \qquad i \in M_{h,g}, \tag{4.3.12a}$$

$$\chi^{i}(w(h,g),h,g) \leqslant 0 \qquad i \in R_{h,g} \setminus M_{h,g}, \tag{4.3.12b}$$

$$\nu^{i}(h,g) = 0 \qquad i \in I \setminus R_{h,g}. \tag{4.3.12c}$$

By Proposition 4.1 for $\alpha > 0$ sufficiently small we have

$$\left| \frac{1}{\alpha} \left[v(h+\alpha g,g)-v(h,g) \right] \right| \leqslant c \;, \quad \left| \frac{1}{\alpha} \left[\mu(h+\alpha g,g)-\mu(h,g) \right] \right| \leqslant c.$$

Hence for any sequence $\{\alpha\}\downarrow 0$ the sequences $\{\frac{1}{\alpha}[v(h+\alpha g,g)-v(h,g)]\}$ and $\{\frac{1}{\alpha} [\mu(h+\alpha g,g)-\mu(h,g)]\}$ have cluster points.

It will be shown that each pair of these cluster points satisfies (4.3.11) and (4.3.12). Hence they are respectively the solution and the associated Lagrange multiplier for $(QP^{(2)}_{h,g})$.

The theorem will follows by the uniqueness of the solution and the multiplier.

Let $\{\alpha\}\downarrow 0$ be any sequence, such that

$$\lim_{\alpha\downarrow 0} \frac{1}{\alpha}\left[v(h+\alpha g,g)-v(h,g)\right] = w \;, \quad \lim_{\alpha\downarrow 0} \frac{1}{\alpha}\left[\mu(h+\alpha g,g)-\mu(h,g)\right]=\nu. \tag{4.3.13}$$

Let us take the difference quotient of (4.1.9a) at $\{h+\alpha g\}$ and at h. Passing to the limit and taking into account definitions (4.3.8) we find that w and ν satisfy (4.3.11a).

Let us verify the remaining conditions (4.3.11) and (4.3.12). First consider the set $M_{h,g}$. By (4.3.5) together with (4.1.9b) and (4.2.2) for $\alpha \in (0,\alpha^{i}]$ we get

$$\psi^{i}(v(h+\alpha g),h+\alpha g,g) = 0. \tag{4.3.14}$$

Taking the difference quotient of (4.3.14) at $(h+\alpha g)$ and at h, passing to the limit and using definitions (4.1.7) and (4.3.9a) we find that w satisfies (4.3.12a).

Now, by (4.1.8), (4.2.9) as well as by (4.1.7) and (4.3.6) for $\alpha \geqslant 0$ and sufficiently small

$$\mu^{i}(h+\alpha g) = 0 \qquad i \in I \setminus R_{h,g}.$$

By (4.3.13) this implies (4.3.12c).

To complete the proof of the theorem it remains to show that on the set $R_{h,g}\setminus M_{h,g}$ conditions (4.3.11b), (4.3.11c) and (4.3.12b) hold.

To this end it is enough to consider the following four cases of sequences $\{\alpha\}\downarrow 0$:

(1) $\psi^i(v(h+\alpha g),h+\alpha g,g) < 0\ ,\quad \mu^i(h+\alpha g,g) = 0,$

(2) $\psi^i(v(h+\alpha g),h+\alpha g,g) > 0\ ,\quad \mu^i(h+\alpha g,g) = 0,$

(3) $\psi^i(v(h+\alpha g),h+\alpha g,g) = 0\ ,\quad \mu^i(h+\alpha g,g) \geq 0,$

(4) $\psi^i(v(h+\alpha g),h+\alpha g,g) = 0\ ,\quad \mu^i(h+\alpha g,g) < 0,$

where $i\in R_{h,g}\setminus M_{h,g}$.

Since $\psi^i(v(h),h,g)=0$, $\mu^i(h,g)=0$ for $i\in R_{h,g}\setminus M_{h,g}$, it is easy to see that in cases (1) and (3)

$$\chi^i(w,h,g) \leq 0\ ,\ \nu^i \geq 0 \quad\text{and}\quad \nu^i\psi^i(w,h,g) = 0, \tag{4.3.15}$$

i.e. conditions (4.3.11b), (4.3.11c) and (4.3.12b) hold.

In case (2) we get

$$\chi^i(w,h,g) \geq 0 \quad\text{and}\quad \nu^i = 0.$$

If $\chi^i(w,h,g)=0$, then the required conditions are satisfied, however if

$$\chi^i(w,h,g) > 0, \tag{4.3.16}$$

then (4.3.12b) is violated.

Finally in case (4) we have

$$\chi^i(w,h,g) = 0 \quad\text{and}\quad \nu^i \leq 0$$

and if $\nu^i=0$, then the required conditions are satisfied, while if

$$\nu^i < 0 \tag{4.3.17}$$

(4.3.11c) is violated.

We shall show that neither (4.3.16) nor (4.3.17) can take place for any $i\in R_{h,g}\setminus M_{h,g}$.

Assume the opposite, namely that for some $j\subset R_{h,g}\setminus M_{h,g}$ there exist three different sequences $\{\alpha\}$, $\{\bar{\alpha}\}$ and $\{\bar{\bar{\alpha}}\}$, for which respectively conditions (4.3.15), (4.3.16) and (4.3.17) are satisfied.

In case of (4.3.15) w is given by the solution of $(QP^{(2)}_{h,g})$.

It is easy to see that in case of (4.3.16) $\bar{w}$ is given by a unique solution of the problem $(\overline{QP}^{(2)}_{h,g})$, which is the same as $(QP^{(2)}_{h,g})$ but with no constraints imposed for index j.

Finally in case of (4.3.17) $\overline{\overline{w}}$ is given by a solution of the problem $(\overline{\overline{QP}}^{(2)}_{h,g})$, which is formulated as $(QP^{(2)}_{h,g})$ with the exception that the equality type constraints in (4.3.9) hold for indices $M_{h,g} \cup \{j\}$. By (A6) the admissible set for $(\overline{\overline{QP}}^{(2)}_{h,g})$ is non-empty, hence there exists a unique solution to this problem. By (4.3.16) and (4.3.17) we have $w \neq \overline{w} \neq \overline{\overline{w}}$.

Note that for all sequences $\{\alpha\}\downarrow 0$ such that convergence (4.3.13) holds, the limit elements in (4.3.13) must satisfy one of the conditions (4.3.15) through (4.3.17). So if the sequences $\{\alpha\}$, $\{\overline{\alpha}\}$ and $\{\overline{\overline{\alpha}}\}$ existed, then the sequences $\{1/\alpha[v(h+\alpha g,g)-v(h,g)]\}$ would have three isolated cluster points. It is impossible by Lemma 4.2. Hence for all sequences $\{\alpha\}\downarrow 0$ the limits (4.3.13) must be the same.

Note that by the definitions (4.2.2), (4.3.5) and (4.3.6) it follows that for any $i \in R_{h,g} \setminus M_{h,g}$ there must exist a sequence $\{\alpha\}\downarrow 0$ such that $\psi^i(v(h+\alpha g),h+\alpha g,g) \leq 0$ and $\mu^i(h+\alpha g,g) \geq 0$, hence for this sequence conditions (4.3.15) hold. Therefore conditions (4.3.16) and (4.3.17) are excluded and the proof of the theorem is completed. □

Provided that the data of Problem (P_h) are sufficiently regular we can repeat the argument of the proof of Theorem 4.3 and thus we obtain

<u>Corollary 4.1</u>

If conditions (A1')-(A4') and (A5), (A6) are satisfied then u(.) and $\lambda(.)$ are k-times right-differentiable at any $h \in H$ in any direction $g \in R^m$, $|g|=1$. The respective right differentials are given by the solution and the associated Lagrange multiplier of an auxiliary quadratic programming problem which can be derived in the same way as $(QP^{(2)}_{h,g})$.

<u>Remark 4.1</u>

In definition (4.3.4) both the first and the second differentiation were performed along the same direction g. The proof of Theorem 4.3 can not be directly extended to the case where these directions of differentiation are different.

In the next section it will be shown that in general right and left differentials do not coincide. Hence the directional differential is a discontinuous function of direction of differentiation and we can not expect higher order differentiability of u(.) and $\lambda(.)$ in arbitrary different directions. The problem of characterizing of those (different) directions in which higher order differentials exist remains open.

It is easy to see that in a particular case where the strict complementarily condition holds at $u(h)$, i.e. if $I_h = I_h^c$, then under assumptions (A1')-(A4'), (A5) and (A6) $u(.)$ and $\lambda(.)$ are k-times Gâteaux differentiable at h in any arbitrary directions g_i, $i=1,2,...,k$.

4.4 Continuous Differentiability

In this Section we are going to derive conditions under which the solutions to (P_h), as well as the associated Lagrange multipliers, are continuously Gâteaux differentiable functions of the parameter h (see [39]).

Using exactly the same argument as in the proof of Theorem 4.1 we find that the left-differential of $u(.)$ at the point h in the direction g

$$v^-(h,g) \overset{\text{def}}{=} \delta_h^- u(h;g) = \lim_{\alpha \downarrow 0} \frac{1}{\alpha} \left[u(h+\alpha g) - u(h) \right] \tag{4.4.1}$$

exists and is given by the solution of the following quadratic programming problem

$$(QP_{h,g}^-) \quad \left| \begin{array}{l} \text{find } v^-(h,g) \in R^n \text{ such that} \\ \\ k(v^-(h,g);h,g) = \min\limits_{v \in \psi_{h,g}^-} k(v;h,g), \end{array} \right. \tag{4.4.2}$$

where $k(v,h,g)$ is given by (4.1.6) and

$$\psi_{h,g}^- = \{ v \in R^n \mid \psi^i(v;h,g) \begin{cases} = 0 & \text{for } i \in I_h^c \\ \geq 0 & \text{for } i \in I_h \setminus I_h^c \end{cases} \}. \tag{4.4.3}$$

The left-differential of Lagrange multipliers $\lambda^i(.)$

$$\mu^{-i}(h;g) \overset{\text{def}}{=} \delta_h^i \lambda^i(h,g) = \lim_{\alpha \downarrow 0} \frac{1}{\alpha} \left[\lambda^i(h+\alpha g) - \lambda^i(h) \right] \tag{4.4.4}$$

for $i \in I_h$ are given as the Lagrange multipliers associated with $(QP_{h,g}^-)$, while

$$\mu^{-i}(h;g) = 0 \quad \text{for} \quad i \in I \setminus I_h. \tag{4.4.4a}$$

In general the solutions to $(QP_{h,g})$ and $(QP_{h,g}^-)$ are different, hence $u(.)$ is not continuously Gâteaux differentiable at h in the direction g.

Continuous differentiability takes place if and only if

$$v(h) = v^{-}(h). \qquad (4.4.5)$$

Taking into account conditions of optimality (4.1.9) for $(QP_{h,g})$, and analogous conditions for $(QP^{-}_{h,g})$ we find that (4.4.5) holds if and only if

$$< D_u\phi^i(u(h),h),v\ (h,g)> + < D_u\phi^i(u(h),h),g> = \qquad (4.4.6a)$$

$$= < D_u\phi^i(u(h),h),v^{-}(h,g)> + < D_u\phi^i(u(h),h),g> = 0 \quad i \in I_h \setminus I_h^c ,$$

$$\mu^i(h) = \mu^{-i}(h) = 0 \qquad i \in I_h \setminus I_h^c . \qquad (4.4.6b)$$

It is easy to see that if condition (4.4.6) are satisfied then also $\lambda(.)$ is continuously Gâteaux differentiable at h in the direction g.

Note that if (4.4.6b) holds, then the inequality type constraints ϕ^i $(i \in I_h \setminus I_h^c)$ do not influence the solutions to $(QP_{h,g})$ and $(QP^{-}_{h,g})$. Hence, in this case the continuous Gâteaux differential $v(h,g) \overset{def}{=} du(h;g)$ of $u(.)$ at h in the direction g is given as the solution of the following quadratic problem of optimization, subject to equality type constraints only:

$$(QP^o_{h,g}) \quad \left| \begin{array}{l} \text{find } v(h,g) \quad R^n \text{ such that} \\ \\ k(v(h,g),h,g) = \min\limits_{v \in \psi^o_{h,g}(h)} k(v,h,g), \end{array} \right. \qquad (4.4.7)$$

where

$$\psi^o_{h,g} = \{v \in R^n \mid <D_u\phi^i(u(h),h),v> + < D_h\phi^i(u(h),h),g> = 0,\ i \in I_h^c\}. \qquad (4.4.8)$$

We shall find conditions under which (4.4.6) holds.
Taking into account (4.1.7), (4.4.3) and (4.4.6a) we obtain

$$D_u\phi_{I_h}(u(h),h)v(h,g) + D_h\phi_{I_h}(u(h),h)g = 0. \qquad (4.4.9)$$

On the other hand from (4.1.9a) and (4.1.10c) we get

$$v(h,g) = -Q^{-1}(h)D_u\phi^T_{I_h}(u(h),h)\mu_{I_h}(h,g) - Q^{-1}(h)q(h,g), \qquad (4.4.10)$$

where, like in Chapter 1, ϕ_{I_h} and μ_{I_h} denote the subvectors of ϕ

and μ respectively containing all components ϕ^i and μ^i, such that $i \in I_h$.

Substituting (4.4.10) into (4.4.9) and taking into account (4.1.6b) yields

$$R(h)\mu_{I_h}(h,g) = S(h)g, \tag{4.4.11}$$

where

$$R(h) = D_u\phi_{I_h}(u(h),h)Q^{-1}(h)D_u\phi_{I_h}^T(u(h),h), \tag{4.4.12a}$$

$$S(h) = -D_u\phi_{I_h}(u(h),h)Q^{-1}(h)\Big[D_{uh}^2 f(u(h),h) + \\ + \sum_{i \in I} \lambda^i(h)D_{uh}^2\phi^i(u(h),h)\Big] + D_h\phi_{I_h}(u(h),h). \tag{4.4.12b}$$

Note that by (4.1.6a) together with (A1), (A3), (A6) and (1.1.16) the matrix $R(h)$ is positive definite, hence (4.4.11) has a unique solution $\mu_{I_h}(h,g)$.

Therefore it is easy to see that (4.4.6b) holds if and only if

$$S(h)g \in \mathrm{lin}\,\{r^i(h) \mid i \in I_h^c\}, \tag{4.4.13}$$

where $r^i(h)$-are the columns of the matrix $R(h)$, while $\mathrm{lin}\{r^i\}$ - denotes the linear span of the set $\{r^i\}$.

Note that (4.4.13) is trivially satisfied if $I_h = I_h^c$, i.e. if the strict complementarity condition holds at $u(h)$. This sufficient condition of continuous differentiability is known in a much more general case (see [16]).

Summing up the above results we arrive at

Theorem 4.4

If conditions (A1) through (A6) are satisfied then $u(.)$ is continuously Gâteaux differentiable at h in the direction g if and only if (4.4.13) holds. The differential $v(h,g) \overset{\text{def}}{=} du(h,g)$ is given by the solution of $(QP_{h,g}^o)$.

In that case also the $\lambda^i(.)$ are differentiable in the same sense and for $i \in I_h^c$ the respective differentials $\mu^i(h,g) \overset{\text{def}}{=} d\lambda^i(h,g)$ are given by the Lagrange multipliers associated with $(QP_{h,g}^o)$, while

$$\mu^i(h,g) = 0 \quad \text{for} \quad i \in I \setminus I_h^c. \tag{4.4.14}$$

Corollary 4.2

If conditions (A1) through (A6) are satisfied then the functions $u(.)$ and $\lambda(.)$ are continuously Gâteaux differentiable (in every direction) if and only if (4.4.13) holds for every direction $g \in R^m$.

Let us consider the so called optimal value function

$$f^o(.) : H \to R^1 ,$$

$$f^o(h) \overset{def}{=} f(u(h),h), \tag{4.4.15}$$

which to every $h \in H$ assigns the corresponding optimal value of the cost functional for (P_h).

It can be proved (see [30]) that $f^o(.)$ is a continuously differentiable function without assumption (4.4.13).

To show this fact first let us note that, since by (1.1.15) the second term in the Lagrangian $L(u,\lambda;h)$ defined by (1.1.12) vanishes at $(u(h),\lambda(h),h)$ we have

$$f^o(h) = L(u(h),\lambda(h),h). \tag{4.4.16}$$

Hence

$$\delta_h^+ f^o(h,g) = \langle D_u L(u(h),\lambda(h),h),v(h,g)\rangle +$$

$$+ \langle D_\lambda L(u(h),\lambda(h),h),\mu(h,g)\rangle + \langle D_h L(u(h),\lambda(h),h),g\rangle . \tag{4.4.17}$$

Note that by (1.1.12), (4.1.3a) and (4.1.10c) we get

$$\langle D_\lambda L(u(h),\lambda(h),h),\mu(h,g)\rangle = \langle \phi(u(h),h),\mu(h,g)\rangle = 0. \tag{4.4.18}$$

Combining (1.1.14), (4.1.17) and (4.1.18) we obtain

$$\delta_h^+ f^o(h,g) = \langle D_h L(u(h),\lambda(h),h),g\rangle . \tag{4.4.19a}$$

Similarly

$$\delta_h^- f^o(h,g) = \langle D_h L(u(h),\lambda(h),h),g\rangle . \tag{4.4.19b}$$

Note that by (A2), (A4) as well as by Theorem 1.1 we find that

$$D_h L(u(h),\lambda(h),h) = D_h f(u(h),h) + D_h \phi^T(u(h),h)\ \lambda(h)$$

is a continuous function of h.

Hence, from (4.4.19) we obtain

Proposition 4.2

If conditions (A1) through (A6) hold, then the optimal value function $f^o(.)$ for (P_h) is continuously Fréchet differentiable, and

$$D_h f^o(h) = D_h L(u(h),\lambda(h),h). \tag{4.4.20}$$

Note that the result (4.4.20) is well known [9] in sensitivity analysis of mathematical programming problems in much more general cases.

4.5 Differentiability of Metric Projection onto Convex Set. Clarke's Generalized Derivative

The results of Section 4.4 will be specialized here to the problem of the metric projection onto a closed and convex set, which is a particular case of (P_h).

For solutions of this problem the form of the Clarke's generalized derivatives,with respect to the parameter h,will be obtained.

If we substitute in Problem (P_h)

$$m = n,\ H = R^n,$$

$$f(u,h) = \frac{1}{2}< u-h,u-h >, \tag{4.5.1}$$

$$\phi^i(u,h)=\phi^i(u),\ i \in I, \text{ are independent of } h,$$

then (P_h) becomes the problem of the metric projection onto a closed and convex set $\phi \in R^n$, considered among others in [29].

In this case by (4.1.6), (4.4.8) and (4.4.12) we get

$$Q(h) = E + \sum_{i \in I_h} \lambda^i(h) D^2_{uu}\phi^i(u(h)), \tag{4.5.2}$$

$$q(h,g) = -g, \tag{4.5.3}$$

$$\Psi^o_{h,g} = \Psi^o_h = \{v \in R^n \mid < D_u\phi^i(u(h)),v >= 0,\ i \in I^c_h\}, \tag{4.5.4}$$

$$S(h) = D_u\phi_{I_h}(u(h))Q^{-1}(h) \tag{4.5.5}$$

where E denotes the unit matrix.

Note that by (A1) and (A6) the matrix S(h) is of the full rank. Since the matrix R(h) is non-singular, then for any $\mu_{I_h}(h,g)$ there exists a direction g such that (4.4.11) holds. This implies that if $I^c_h \neq I_h$, then there always exist directions $g \in R^n$ for which (4.4.6b) is violated and $u(.)$ is not continuously Gâteaux differentiable at h in the direction g.

Thus we proved that a necessary condition for continuous Gâteaux differentiability of $u(.)$ at h is that the strict complementarity

$$I_h^c = I_h \tag{4.5.6}$$

holds.

Of course (4.5.6) is also a sufficient condition for continuous Gâteaux differentiability and the differential in any direction g is given by the solution of $(QP^o_{h,g})$. The same holds for differentiability of $\lambda(.)$.

It is easy to see that (4.5.6) implies more, namely Fréchet differentiability of $u(.)$ and $\lambda(.)$ at h.

Indeed if (4.5.6) holds then by Theorem 1.3 there exists a neighbourhood $X \subset H$ of h, such that for each $x \in X$ we have

$$I_x^c = I_x = I_h^c = I_h.$$

Hence the function $u(.)$ is continuously Gâteaux differentiable at each $x \in X$ and the differential is characterized by the solution of (QP^o_x) with the same set of constraints

$$\psi_h^o = \psi_x^o \qquad \forall x \in X. \tag{4.5.7}$$

Of course, if (4.5.7) holds then the solution to (QP^o_x) and the associated multipliers are continuous functions of x on X.

This completes the proof of the Fréchet differentiability.

Thus, we proved the following:

Theorem 4.5

If condition (A6) holds, then the metric projection $u(.)$ onto Φ and the associated Lagrange multipliers $\lambda(.)$ are continuously Fréchet differentiable at h if and only if the strict complementarity condition (4.5.6) holds. The respective differentials are characterized by the solution and the associated multipliers of $(QP^o_{h,g})$ as well as by (4.4.14).

Now let us recall the notion of the generalized derivative in Clarke's sense [7,9]

Definition 4.1

If $f(.)$ is a Lipschitz continuous function defined on a neighbourhood of $h \in R^n$, then the generalized, in Clarke's sense, derivative of $f(.)$ at h is given by

$$\partial f(h) = \text{conv}\{L \mid \exists \{h_i\} \to h \text{ with } f(.) \text{ Fréchet differentiable}$$
$$\text{at } h_i \text{ and } Du(h_i) \to L\}. \tag{4.5.8}$$

We are going to characterize the generalized derivatives, in Clarke's sense, of $u(.)$ and $\lambda(.)$ at those points h at which these functions are not Fréchet differentiable.

Let $h \in H$ be any arbitrary point at which $u(.)$ is not continuously differentiable, i.e. by Theorem 4.5

$$I_h^c \neq I_h.$$

Note that by Theorem 1.3 and by Rademacher's theorem the functions $u(.)$ and $\lambda(.)$ are Fréchet differentiable almost everywhere on H. Hence the point h can be approached by a sequence $\{x\} \subset H$ of points x at which $u(.)$ and $\lambda(.)$ are differentiable, i.e. by Theorem 4.5

$$I_x^c = I_x.$$

By Theorem 1.3 at the points x close enough to h we have

$$I_h^c \subset I_x^c = I_x \subset I_h. \tag{4.5.9}$$

The differentials of $u(.)$ at the points $x \in \{x\}$ in the direction g are given by the solutions to $(QP_{x,g}^o)$, where $I_x^c = I_x$ satisfies (4.5.9).

<u>Lemma 4.3</u>

If condition (A6) holds, then for any set of indices I_x satisfying (4.5.9) there exist a direction $g \in R^n$ and a sequence $\{y\} \subset H$ converging to h along the direction (-g) such that $I_y^c = I_y = I_x$.

<u>Proof</u>

Since

$$\phi^i(u(h)) = 0 \quad \text{and} \quad \lambda^i(h) = 0 \quad \text{for} \quad i \in I_h \setminus I_h^c,$$

then, taking into account (1.1.15) it is easy to see that to prove the lemma it is enough to show that for any I_x satisfying (4.5.9) there exists a direction g, such that $v(h,g)$ and $\mu(h,g)$ satisfy the conditions;

$$\langle D_u\phi^i(u(h)), v(h,g) \rangle < 0 \qquad \text{for} \quad i \in I_h \setminus I_x, \tag{4.5.10a}$$

$$\mu^i(h,g) > 0 \qquad \text{for} \quad i \in I_x \setminus I_h^c. \tag{4.5.10b}$$

We shall show that there exists a direction g for which (4.5.10) holds. Namely we shall construct a direction g such that

$$< D_u\phi^i(u(h)), v(h,g) > = \begin{cases} 0 & \text{for} \quad i \in I_x \setminus I_h^c \\ -1 & \text{for} \quad i \in I_h \setminus I_x \, , \end{cases} \tag{4.5.11a}$$

$$\mu_h^i = \begin{cases} 1 & \text{for} \quad i \in I_x \setminus I_h^c \\ 0 & \text{for} \quad i \in I_h^c \ \text{and} \ i \subset I_h \setminus I_x \, . \end{cases} \tag{4.5.11b}$$

Assume that (4.5.11) holds, then from (4.5.4) and (4.5.11a) we obtain

$$D_u\phi_{I_h}(u(h))v(h,g) = -\rho_x \, , \tag{4.5.12}$$

where ρ_x is the vector of appropriate dimension with components

$$\rho_x^i = \begin{cases} 1 & \text{for} \quad i \in I_h \setminus I_x \\ 0 & \text{for} \quad i \in I_x. \end{cases} \tag{4.5.12a}$$

Substituting (4.4.10) into (4.5.12) and using (4.4.12a), (4.5.3) and (4.5.5) we obtain

$$S(h)g = R(h)\mu_{I_h}(h) - \rho_x \, . \tag{4.5.13}$$

Since the matrix $S(h)$ is of the full rank, then for any $\mu_{I_h}(h)$ and any ρ_x there exists a direction g, such that (4.5.13) holds. □

By Theorem 4.5, for each $y \in \{y\}$ and for any $g \in R^n$ the differential $(D_h u(y), g)$ is given by the solution $v(y,g)$ of $(QP^o_{y,g})$, where $I_y^c = I_x$.

It is obvious that

$$\lim_{y \to h} v(y,g) = v_x(h,g), \tag{4.5.14}$$

where $v_x(h,g)$ is the solution of the following problem of optimization

$$(QP^x_{h,g}) \quad \left| \begin{array}{l} \text{find } v_x(h,g) \in R^n, \text{ such that} \\ k(v_x(h,g);h,g) = \min\limits_{v \in \Psi_h^x} k(v,h,g), \end{array} \right. \tag{4.5.15}$$

where

$$\Psi_h^x = \{ v \in R^n | < D_u\phi^i(u(h)), v > = 0 \, , \quad i \in I_x \}. \tag{4.5.15a}$$

Similarly

$$\lim_{y \to h} \mu^i(y,g) = \mu^i_x(h,g), \tag{4.5.16}$$

where $\mu^i_x(h,g)$, $i \in I_x$, are Lagrange multipliers associated with $(QP^x_{h,g})$, while

$$\mu^i_x(h,g) = 0 \ , \quad i \in I \setminus I_x \ . \tag{4.5.17}$$

From Definition 4.1, Lemma 4.3 and from (4.5.14), (4.5.16) we obtain

Theorem 4.6

If condition (A6) holds, then the generalized differentials in Clarke's sense of the metric projection $u(.)$ onto Φ and of the associated Lagrange multipliers $\lambda(.)$ at h in the direction g are given by

$$\begin{aligned} \partial u(h)g &= \text{conv}\{v_x(h,g) \quad \text{for all} \quad I^c_h \subset I_x \subset I_h\}, \\ \partial \lambda(h)g &= \text{conv}\{\mu_x(h,g) \quad \text{for all} \quad I^c_h \subset I_x \subset I_h\}, \end{aligned} \tag{4.5.18}$$

where $v_x(h,g)$ is the solutions to $(QP^x_{h,g})$, while $\mu_x(h,g)$ is given by the associated Lagrange multipliers and by (4.5.17).

We have to stress here, that in a general case of convex programming (P_h) Theorem 4.6 may not be true, i.e.

$$\partial u(h)g \neq \text{conv}\{v_x(h,g) \quad \text{for all} \quad I^c_h \subset I_x \subset I_h\}.$$

To show this fact let us consider the following simple example.

Let $n=2$, $m=1$, $H=R^1$ and

$$f(u,h) = \langle u-hw, u-hw \rangle ,$$

$$\phi^i(u,h) = \langle a^i, u \rangle \qquad i=1,2,$$

where

$$w = (1,0)^T, \ a^1 = (1,1)^T, \ a^2 = (-1,1)^T$$

are fixed vectors independent of h.

For these data we are looking for the Clarke's generalized derivative of the solutions $u(.)$ to (P_h) at $h=0$.

If $h \neq 0$ then we have

$$u(h) = \frac{h}{2}(1,-1)^T \qquad \text{for} \quad h > 0,$$

$$u(h) = \frac{h}{2}(1,1)^T \qquad \text{for} \quad h < 0.$$

Hence we get respectively

$$D_h\, u(h) = \frac{1}{2}\,(1,-1)^T \qquad \text{for} \quad h > 0,$$

$$D_h\, u(h) = \frac{1}{2}\,(1,1)^T \qquad \text{for} \quad h < 0,$$

and by definition

$$\partial u(0) = \text{conv}\,\{\, \frac{1}{2}\,(1,1)^T,\ \frac{1}{2}(1,-1)^T\}. \qquad (4.5.19)$$

Now let us try to apply Theorem 4.6.
To this end we must find the solutions to $(QP^x_{h,g})$. It is easy to see that

$$Q(h) = E \quad , \quad q(h,g) = -w,$$

$$\Psi^x_{h,g} = \{v \in R^n \mid \langle a^i, v\rangle = 0,\ i \in I_x\},$$

where according to (4.5.9)

$$I_x = \{1\},\ \{2\},\ \{1,2\}.$$

For these three possible sets of constraints we obtain the following solutions to $(QP^x_{h,g})$

$$v_x(0) = \frac{1}{2}\,(1,-1)^T,\ \frac{1}{2}\,(1,1)^T,\ (0,0)^T.$$

Therefore

$$\text{conv}\{v_x(0) \quad \text{for all} \quad I^c_o \subset I_x \subset I_o\} =$$

$$= \text{conv}\,\{\,\frac{1}{2}\,(1,-1)^T,\ \frac{1}{2}\,(1,1)^T,\ (0,0)^T\}. \qquad (4.5.20)$$

Comparing (4.5.19) and (4.5.20) we find that

$$\partial u(0) \subset \text{conv}\{v_x(0) \quad \text{for all} \quad I^c_o \subset I_x \subset I_o\} \quad \text{and}$$

$$\partial u(0) \neq \text{conv}\{v_x(0) \quad \text{for all} \quad I^c_o \subset I_x \subset I_o\}.$$

Hence, for the considered example the result analogous to Theorem 4.6 does not hold.

5. DIFFERENTIAL STABILITY OF SOLUTIONS TO OPTIMAL CONTROL PROBLEMS FOR DISCRETE SYSTEMS

In this chapter we shall investigate differential properties of solutions to convex optimal control problems for systems described by linear difference equations (see [40]). The optimal control problems will be reformulated as convex programming problems and the results of Chapter 4 will be used to obtain the form of the right-differentials of the solutions to these problems.

5.1. Problem Statement

Like in the previous chapter $H \subset R^m$ denotes an open and convex set of vector parameters.

For each $h \in H$ we define the following convex optimal control problem (DC_h) for a linear discrete system

(DC_h)

find a pair $(u(h),\ x(h))$, where

$$u(h) = [u_o^T(h), u_1^T(h), \dots, u_{k-1}^T(h)]^T \in R^{n.k},$$

$$x(h) = [x_o(h), x_1(h), \dots, x_k(h)]^T \in R^{\ell(k+1)}, \text{ such that}$$

$$F(u(h),x(h),h) = \min\{F(u,x,h) \overset{\text{def}}{=} \sum_{j=0}^{k-1} f_j(u_j,x_j,h)\}, \tag{5.1.1}$$

subject to

$$\nabla x_j = A_j(h)x_j + B_j(h)u_j, \quad j=0,1,\dots,k-1, \tag{5.1.2}$$

$$x_o = 0, \tag{5.1.2a}$$

$$\phi_j^i(u_j,h) \leqslant 0, \quad j=0,1,\dots,k-1, \quad i=1,2,\dots,r, \tag{5.1.3a}$$

$$\theta_j^i(x_j,h) \leqslant 0, \quad j=0,1,\dots,k, \quad i=1,2,\dots,s, \tag{5.1.3b}$$

where

$$\nabla x_j \overset{\text{def}}{=} x_{j+1} - x_j, \quad A_j(\cdot): R^m \to R^{\ell x \ell}, \quad B_j(\cdot): R^m \to R^{\ell x n},$$

$$f_j(\cdot,\cdot,\cdot): R^n \times R^\ell \times R^m \to R^1, \quad \phi_j^i(\cdot,\cdot): R^n \times R^m \to R^1, \quad \theta_j^i(\cdot,\cdot): R^\ell \times R^m \to R^1.$$

Denote

$$\phi_j(u_j,h) = [\phi_j^1(u_j,h), \phi_j^2(u_j,h), \dots, \phi_j^r(u_j,h)]^T,$$

$$\phi(u,h) = [\phi_o^T(u_o,h), \phi_1^T(u_1,h), \dots, \phi_{k-1}^T(u_2,h)]^T,$$

$$\theta_j(x_j,h) = [\theta_j^1(x_j,h), \theta_j^2(x_j,h), \dots, \theta_j^s(x_j,h)]^T,$$

$$\theta(x,h) = [\theta_o^T(x_o,h), \theta_1^T(x_1,h), \dots, \theta_k^T(x_k,h)]^T.$$

Assume that the following conditions hold:

(D1) for each $h \in H$ the functions $f_j(.,.,h)$ $(j=0,1,\dots,k-1)$ are twice continuously differentiable in both arguments. Moreover $f_j(.,.,h)$ are strongly convex, uniformly with respect to h, i.e. there exists a constant $\alpha > 0$, independent of h, such that

$$[v^T, y^T] \begin{bmatrix} D^2_{uu} f_j(u,x,h), & D^2_{ux} f_j(u,x,h) \\ D^2_{xu} f_j(u,x,h), & D^2_{xx} f_j(u,x,h) \end{bmatrix} \begin{bmatrix} v \\ y \end{bmatrix} \geq \alpha(|v|^2 + |y|^2), \qquad (5.1.4)$$

$\forall u,v \in R^n$, $\forall x,y \in R^n$, $\forall h \in H$, $\forall j=0,1,\dots,k-1$,

(D2) the functions $f_j(.,.,.)$, $D_u f_j(.,.,.)$ and $D_x f_j(.,.,.)$ $(j=0,1,\dots,k-1)$ are continuously differentiable in all variables,

(D3) $A_j(.)$ and $B_j(.)$ $(j=0,1,\dots,k-1)$ are continuously differentiable,

(D4) for each $h \in H$ the functions $\phi_j(.,h)$ $(j=0,1,\dots,k-1)$ and $\theta_j(.,h)$ $(j=0,1,\dots,k)$ are twice continuously differentiable and convex,

(D5) the functions $\phi_j(.,.)$, $D_u\phi_j(.,.)$ $(j=0,1,\dots,k-1)$ and $\theta_j(.,.)$, $D_x\theta_j(.,.)$ $(j=0,1,\dots,k)$ are continuously differentiable in all variables,

(D6) for each $h \in H$

$$\theta_o(x_o,h) < 0, \qquad (5.1.5)$$

(D7) for each $h \in H$ there exists a pair $(\hat{u}(h), \hat{x}(h))$ satisfying (5.1.2) and (5.1.3).

Note that by (D1), (D4) and (D7) Problem (DC_h) has a unique solution. In addition we assume:

(D8) there exists a constant $\beta > 0$, independent of h, such that

$$|[\Omega_j(h), B^T(h)\Lambda_{j+1}(h)]v| \geq \beta|v| \qquad j=0,1,\dots,k-1 \qquad (5.1.6)$$

for every $h \in H$ and for every v of appropriate dimension, where

$$\Omega_j(h) \stackrel{def}{=} D_u\phi^T_{I^j_{hu}}(u_j(h),h) , \quad \Lambda_j(h) \stackrel{def}{=} D_x\theta^T_{I^j_{hx}}(x_j(h),h) \qquad (5.1.6a)$$

are the matrices whose columns are the gradients of all functions

$\phi_j^i(\cdot,h)$, $i=1,2,\ldots,r$, and $\theta_j^i(\cdot,h)$, $i=1,2,\ldots,s$, binding at $u_j(h)$ and $x_j(h)$ respectively.

5.2. Right-Differentiability of Solutions

Our purpose is to find the form of the right-differentials with respect to h of the solutions $u(h)$, $x(h)$ to (DC_h) as well as of the associated Lagrange multipliers.

To this end, in a standard way (see [6]), we reformulate (DC_h) as a convex programming problem in $R^{n.k} \times R^{\ell(k+1)}$. Namely denoting $w \overset{\text{def}}{=} (u^T,x^T)^T$ we rewrite (DC_h) as:

(DC_h')

find $w(h) \in R^{n.k} \times R^{\ell.(k+1)}$ such that

$$F(w(h),h) = \min F(w,h), \tag{5.2.1}$$

subject to

$$C(h)w = 0, \tag{5.2.2}$$

$$\Delta(w,h) \leqslant 0, \tag{5.2.3}$$

where $C(h)$ is $[\ell(k+1)]\times[\ell(k+1)+n.k]$ dimensional matrix corresponding to equality type constraints (5.1.2), while $\Delta(w,h)$ is $[rk+s(k+1)]$ - dimensional vector function, containing all inequality type constraints (5.1.3).

By Remarks 1.1 and 4.1 it follows that to (DC_h') we can apply Theorems 1.3 and 4.1 provided that the assumptions (A1) through (A6) hold. It is easy to verify that (D1)-(D7) imply that (A1)-(A5) are satisfied for (DC_h'). Hence, it remains to show that (A6) holds.

Let us construct the matrix $\mathcal{P}_h$, whose columns are the gradients of all the constraints (5.2.2) and all constraints (5.2.3) binding at $w(h)$.

The following lemma shows that for (DC_h') condition (A6) holds:

Lemma 5.1

If conditions (D3), (D5), (D6) and (D8) hold, then for each compact set $\mathcal{H} \subset H$ there exists a constant $\tilde{\beta} > 0$ such that

$$|\mathcal{P}_h\xi| \geqslant \tilde{\beta}\ |\xi| \tag{5.2.4}$$

for all $h \in \mathcal{H}$ and for all vectors ξ of appropriate dimension.

Proof

Constraints (5.2.2) and (5.2.3) are given by (5.1.2) and (5.1.3) respectively, hence taking into account (5.1.5) we can write $\mathcal{P}_h$ in the following form

$$
\mathcal{P}_h = \left[\begin{array}{c|ccc|ccc|c|ccc}
E & 0 & -E-A_o^T(h) & 0 & 0 & 0 & 0 & & & & \\
0 & \Omega_o(h) & -B_o^T(h) & 0 & 0 & 0 & 0 & & & & \\
0 & 0 & E & \Lambda_1(h) & 0 & -E-A_1^T(h) & 0 & & & & \\
 & & & & \Omega_1(h) & -B_1^T(h) & 0 & & & & \\
 & & & & 0 & E & \Omega_2(h) & & & & \\
 & & & & & & & \ddots & & & \\
 & & & & & & & & 0 & -E-A_{k-1}^T(h) & 0 \\
 & & & & & & & & \Omega_{k-1}(h) & -B_{k-1}^T(h) & 0 \\
 & & & & & & & & 0 & E & \Lambda_k(h)
\end{array}\right]
$$

where $\Omega_j(h)$ and $\Lambda_j(h)$ are given by (5.1.6a) and E denotes the unit matrix.

Let $\xi = [\xi_1^T, \xi_2^T, \ldots, \xi_{3k}^T, \xi_{3k+1}^T]^T$ be any arbitrary vector of the dimension equal to the number of columns of the matrix $\mathcal{P}_h$.

ξ_i - are the subvectors of ξ corresponding to the appropriate submatrices of $\mathcal{P}_h$. Denote by $\eta = [\eta_1^T, \eta_2^T, \ldots, \eta_{2k+1}^T]^T$ the vector given by

$$\eta = \mathcal{P}_h \xi \ . \tag{5.2.5}$$

We have to show that there exists a constant $\tilde{\beta}$ independent of $h \in \mathcal{H}$ such that

$$|\eta| \geq \tilde{\beta} |\xi| . \tag{5.2.6}$$

From the last two rows of the vector equation (5.2.5) we get

$$\Omega_{k-1}(h)\ \xi_{3k-1} - B_{k-1}^T(h)\ \xi_{3k} = \eta_{2k} , \tag{5.2.7a}$$

$$\xi_{3k} + \Lambda_k(h)\ \xi_{3k+1} = \eta_{2k+1} . \tag{5.2.7b}$$

Hence

$$\Omega_{k-1}(h)\ \xi_{3k-1} + B_{k-1}^T(h)\ \Lambda_k(h)\ \xi_{3k+1} - B_{k-1}^T(h)\eta_{2k+1} = \eta_{2k} .$$

Taking into account (5.1.6) we obtain

$$\beta(|\xi_{3k-1}|^2+|\xi_{3k+1}|^2)^{\frac{1}{2}} \leqslant |\eta_{2k}|+|B_{k-1}(h)||\eta_{2k+1}| \leqslant$$

$$\leqslant \max\{1,|B_{k-1}(h)|\}\sqrt{2}(|\eta_{2k}|^2+|\eta_{2k+1}|^2)^{\frac{1}{2}}.$$

Hence

$$\beta'(|\xi_{3k-1}|^2+|\xi_{3k+1}|^2)^{\frac{1}{2}} \leqslant (|\eta_{2k}|^2+|\eta_{2k+1}|^2)^{\frac{1}{2}}, \qquad (5.2.8)$$

where

$$\beta' = \frac{\beta}{\sqrt{2}}\min\{1,b\} \quad , \quad b = \min_{\substack{0\leqslant j\leqslant k-1\\ h\in\mathcal{H}}}|B_j(h)|^{-1}. \qquad (5.2.8a)$$

On the other hand from (5.2.7b)

$$|\xi_{3k}|^2 \leqslant 2[|\eta_{2k+1}|^2+|\Lambda_k(h)|^2|\xi_{3k+1}|^2]. \qquad (5.2.9)$$

After elementary evaluations (5.2.8) and (2.5.9) yield

$$\beta''(|\xi_{3k-1}|^2+|\xi_{3k}|^2+|\xi_{3k+1}|^2)^{\frac{1}{2}} \leqslant (|\eta_{2k}|^2+|\eta_{2k+1}|^2)^{\frac{1}{2}}, \qquad (5.2.10)$$

where

$$\beta'' = \beta'(1+2c^2+2\beta'^2)^{-\frac{1}{2}} \quad , \quad c = \max_{\substack{1\leqslant j\leqslant k\\ h\in\mathcal{H}}}|\Lambda_j(h)|. \qquad (5.2.10a)$$

Now from the 3-rd and the 4-th rows from below of (5.2.5) we get equations analogous to (5.2.7)

$$\Omega_{k-2}(h)\xi_{3(k-1)-1}-B^T_{k-2}(h)\xi_{3(k-1)} = \eta_{2(k-1)}, \qquad (5.2.11a)$$

$$\xi_{3(k-1)}+\Lambda_{k-1}(h)\xi_{3(k-1)+1} = \tilde{\eta}_{2(k-1)+1}, \qquad (5.2.11b)$$

where

$$\tilde{\eta}_{2(k-1)+1} = \eta_{2(k-1)+1}(E+A_{k-1}(h))\xi_{3k}. \qquad (5.2.11c)$$

In exactly the same way as in the case of (5.2.10) we get from (5.2.11)

$$\beta''(|\xi_{3(k-1)-1}|^2+|\xi_{3(k-1)}|^2+|\xi_{3(k-1)+1}|^2)^{\frac{1}{2}} \leqslant (|\eta_{2(k-1)}|^2+|\tilde{\eta}_{3(k-1)+1}|^2)^{\frac{1}{2}}. \qquad (5.2.12)$$

(5.2.10) together with (5.2.12) yield

$$\beta''\Big(\sum_{i=3(k-1)-1}^{3k+1}|\xi_i|^2\Big)^{\frac{1}{2}} \leqslant (|\eta_{2k}|^2+|\eta_{2k+1}|^2+|\eta_{2(k-1)}|^2+|\tilde{\eta}_{2(k-1)+1}|^2)^{\frac{1}{2}} \leqslant$$

$$\leqslant \Big(\sum_{i=2(k-1)}^{2k+1}|\eta_i|^2\Big)^{\frac{1}{2}}+a|\xi_{3k}| \leqslant \Big(\sum_{i=2(k-1)}^{2k+1}|\eta_i|^2\Big)^{\frac{1}{2}}+\frac{a}{\beta''}(|\eta_{2k}|^2+|\eta_{2k+1}|^2)^{\frac{1}{2}},$$

where

$$a = \max_{\substack{0\leqslant j\leqslant k-1\\ h\in\mathcal{H}}} |E+A_j^T(h)|. \tag{5.2.12a}$$

Hence

$$\beta'''\Big(\sum_{i=3(k-1)-1}^{3k+1}|\xi_i|^2\Big)^{\frac{1}{2}} \leqslant \Big(\sum_{i=2(k-1)}^{2k+1}|\eta_i|^2\Big)^{\frac{1}{2}},$$

where

$$\beta''' = \beta''\Big(1+\frac{a}{\beta''}\Big)^{-1}.$$

Taking into account the structure of the matrix $\mathcal{P}_h$ and proceeding by induction we finally get

$$\tilde{\beta}|\xi| = \tilde{\beta}\Big(\sum_{i=1}^{3k+1}|\xi_i|^2\Big)^{\frac{1}{2}} \leqslant \Big(\sum_{i=1}^{2k+1}|\eta_i|^2\Big)^{\frac{1}{2}} = |\eta_i|, \tag{5.2.13}$$

where

$$\tilde{\beta} = \beta''\Big(1+\frac{a}{\beta''}\Big)^{-k}. \tag{5.2.13a}$$

□

Lemma 5.1 shows that for (DC_h') the condition (A6) is satisfied. Therefore we can apply to (DC_h') Theorems 1.3 and 4.1.

Theorem 1.3 expressed in terms of the initial Problems (DC_h) takes on the form:

<u>Theorem 5.1</u>

If the assumptions (D1) through (D8) hold, then for each compact and convex set $\mathcal{H} \subset H$ there exists a constant $c > 0$ independent of h such that

$$|u(h_2)-u(h_1)|, |x(h_2)-x(h_1)|, |p(h_2)-p(h_1)|, |\lambda(h_2)-\lambda(h_1)|, |\nu(h_2)-\nu(h_1)| \leqslant$$

$$\leqslant c|h_2-h_1|, \tag{5.2.14}$$

where $p(h)$, $\lambda(h)$ and $\nu(h)$ are the unique Lagrange multipliers associated with the constraints (5.1.2), (5.1.3a) and (5.1.3b) respectively.

Let us recall [6] that $p(h)$ is given as the solution of the following equation adjoint to (5.1.2):

$$\nabla p_j(h) = -A_j^T(h)p_j(h) + D_x f_j(u_j(h),x_j(h),h) + D_x\Theta_j^T(x(h),h)\nu_j(h)$$

$$j=k-1,k-2,\dots,1,0, \qquad (5.2.14)$$

$$p_k(h) = 0. \qquad (5.2.14a)$$

Moreover

$$\nu_k(h) = 0.$$

From Theorem 4.3 we obtain:

Theorem 5.2

If the assumptions (D1) through (D8) hold, then the solutions $(u(\cdot), x(\cdot))$ of (DC_h) and the associated Lagrange multipliers $p(\cdot),\lambda(\cdot)$ and $\nu(\cdot)$ are directionally-differentiable functions of h at any $h \in H$ in any direction $g \in R^m$, $|g|=1$.

The right-differentials $v(h,g) \overset{def}{=} \delta_h^+ u(h,g)$ and $y(h,g)=\delta_h^+ x(h,g)$ are given by the solution of the following quadratic optimal control problem

(QD_h)

find $(v(h,g),y(h,g)) \in R^{n.k} \times R^{\ell(k+1)}$ such that

$$K(v(h,g);y(h,g),h)=\min\{K(v;y,h)=\sum_{j=0}^{k-1} K_j(v;y,h)\}, \qquad (5.2.15)$$

subject to

$$\nabla y_j=A_j(h)y_j+B_j(h)v_j+(D_hA_j(h)g)x_j(h)+(D_hB_j(h)g)u_j(h)),$$

$$j=0,1,\dots,k-1 \qquad (5.2.16)$$

$$y_o = 0, \qquad (5.2.16a)$$

$$\langle D_u\phi_j^i(u_j(h),h),v_j\rangle+\langle D_h\phi_j^i(u_j(h),h),g\rangle \begin{cases} =0 & \text{for } i \in J_{h,u}^j \\ \leq 0 & \text{for } i \in I_{h,u}^j \setminus J_{h,u}^j \end{cases}$$

$$j=0,1,\dots,k-1, \qquad (5.2.17a)$$

$$\langle D_x\Theta_j^i(x_j(h),h),y_j\rangle+\langle D_h\Theta_j^i(x_j(h),h),g\rangle \begin{cases} =0 & \text{for } i \in J_{h,x}^j \\ \leq 0 & \text{for } i \in I_{h,x}^j \setminus J_{h,x}^j \end{cases}$$

$$j=0,1,\dots,k, \qquad (5.2.17b)$$

where

$$K_j(v_j;y_j,h) = \frac{1}{2}[v_j^T, y_j^T]\begin{bmatrix} Q_j^{11}(h), Q_j^{12}(h) \\ Q_j^{21}(h), Q_j^{22}(h)\end{bmatrix} + \begin{bmatrix} v_j \\ y_j\end{bmatrix} + \langle \bar{q}_j(h,g), v_j\rangle + \langle \bar{\bar{q}}_j(h,g), y_j\rangle, \tag{5.2.18}$$

$$Q_j^{11}(h) = D_{uu}^2 f_j(u_j(h), x_j(h), h) + \sum_{i=1}^{r} \lambda_j^i(h) D_{uu}^2 \phi_j^i(u_j(h), h), \tag{5.2.18a}$$

$$Q_j^{22}(h) = D_{xx}^2 f_j(u_j(h), x_j(h), h) + \sum_{i=1}^{s} \nu_j^i(h) D_{xx}^2 \Theta_j^i(x_j(h), h), \tag{5.2.18b}$$

$$Q_j^{12}(h) = \left[Q_j^{21}(h)\right]^T = D_{ux}^2 f_j(u_j(h), x_j(h), h), \tag{5.2.18c}$$

$$\bar{q}_j(h,g) = [D_{uh}^2 f_j(u_j(h), x_j(h), h) + \sum_{i=1}^{r} \lambda_j^i(h) D_{uh}\phi_j^i(u_j(h), h) - D_h B_j^T(h) p_j(h)]g, \tag{5.2.18d}$$

$$\bar{\bar{q}}_j(h,g) = [D_{xh}^2 f_j(u_j(h), x_j(h), h) + \sum_{i=1}^{s} \nu_j^i(h) D_{xh}\Theta_j^i(x_j(h), h) - D_h A_j^T(h) p_j(h)]g, \tag{5.2.18e}$$

$$I_{h,u}^j = \{u \in R^n \mid \phi_j^i(u_j(h), h) = 0, \quad i=1,2,\dots,r\}, \tag{5.2.19a}$$

$$J_{h,u}^j = \{u \in I_{h,u}^j \mid \lambda_j^i(h) > 0\}, \tag{5.2.19b}$$

$$I_{h,x}^j = \{x \in R^\ell \mid \Theta_j^i(x_j(h), h) = 0, \; i=1,2,\dots,s\}, \tag{5.2.20.a}$$

$$J_{h,x}^j = \{x \in I_{h,x}^j \mid \nu_j^i(h) > 0\}. \tag{5.2.20b}$$

The Lagrange multipliers are also right-differentiable and $r_j(h,g) \overset{def}{=} \delta_h^+ p_j(h,g)$, $\mu_j^i(h,g) \overset{def}{=} \delta_h^+ \lambda_j^i(h,g)$, $i \in I_{h,u}^j$, as well as $\pi_j^i(h,g) \overset{def}{=} \delta_h^+ \nu_j^i(h,g)$, $i \in I_{h,x}^j$ are given as the respective multipliers for (QD_h), while

$$\mu_j^i(h,g) = 0 \qquad \text{for} \qquad i \notin I_{h,u}^j, \tag{5.2.21a}$$

$$\pi_j^i(h,g) = 0 \qquad \text{for} \qquad i \notin I_{h,x}^j. \tag{5.2.21b}$$

<u>Remark 5.1</u>

Using the formulation (DC_h') of Problem (DC_h) all stability properties of the convex programming problems, obtained in Chapter 4 can be transfered to the convex optimal control problems for discrete systems.

6. DIFFERENTIAL STABILITY OF SOLUTIONS TO OPTIMAL CONTROL PROBLEMS SUBJECT TO CONTROL CONSTRAINTS

This chapter is devoted to analysis of the differential properties of the convex optimal control problems, subject to pointwise control constraints. In the first two sections the results for the abstract problem formulated in Chapter 2 are derived, while in Section 6.3 and 6.4 the examples of applications to some specific control problems are presented. Note that the abstract approach is quite general and may be applied to a broad class of convex optimal control problems, subject to the pointwise control constraints. For another application see [41].

6.1. Abstract Problem. Right Differentiability

Let us recall Family $\{O_h\}$ of the convex problems of optimization (O_h), depending on a vector parameter $h \in H \subset R^m$, which was defined in Chapter 2:

(O_h)

find a pair $(u_h, z_h) \in U \times Z$ such that

$$F(u_h, z_h, h) = \min_{u \in U_h^{ad}} F(u,z,h), \tag{6.1.1}$$

subject to

$$z = S(h)u, \tag{6.1.2}$$

where

$$U = L^2(\Xi; R^n),$$

$$U_h^{ad} = \{u \in U \mid u(\xi) \in \Phi_h \quad \text{for a.a.} \quad \xi \in \Xi\} \tag{6.1.3}$$

$$\Phi_h = \{u \in R^n \mid \phi^i(u,h) \leqslant 0, \quad i \in I \overset{\text{def}}{=} \{1,2,\dots,r\}\}. \tag{6.1.3a}$$

It is assumed that the conditions (B1) through (B9) of Chapter 2 hold. Hence (O_h) has a unique solutions and moreover the Lagrange multipliers p_h and λ_h associated with the constraints (6.1.2) and (6.1.3) are defined uniquely.

Like in (2.1.10b) we denote

$$I_h(\xi) = \{i \in I \mid \phi^i(u_h(\xi), h) = 0\}, \tag{6.1.4a}$$

and moreover we define

$$I_h^c(\xi) = \{i \in I_h(\xi) \mid \lambda_h(\xi) > 0\}. \tag{6.1.4b}$$

Our purpose is to show that the solutions (u_h, z_h) to (P_h) as well as the associated Lagrange multipliers p_h, λ_h are right-differentiable functions of $h \in H$, i.e. that for any $h \in H$ and for any direction $g \in R^m$, $|g|=1$ there exist

$$\delta_h^+ u(h,g) = \lim_{\alpha \downarrow 0} \frac{1}{\alpha}(u_{h+\alpha g} - u_h), \tag{6.1.5a}$$

$$\delta_h^+ z(h,g) = \lim_{\alpha \downarrow 0} \frac{1}{\alpha}(z_{h+\alpha g} - z_h), \tag{6.1.5b}$$

$$\delta_h^+ p(h,g) = \lim_{\alpha \downarrow 0} \frac{1}{\alpha}(p_{h+\alpha g} - p_h), \tag{6.1.5c}$$

$$\delta_h^+ \lambda(h,g) = \lim_{\alpha \downarrow 0} \frac{1}{\alpha}(\lambda_{h+\alpha g} - \lambda_h), \tag{6.1.5d}$$

where the limits are taken in the strong topologies of respective spaces.

The following theorem was first proved in [41]:

<u>Theorem 6.1</u>

If the assumptions (B1) through (B9) hold, then the solutions (u_h, z_h) of (O_h) and the associated Lagrange multipliers p_h, λ_h are directionally differentiable function of the parameter h, at any $h \in H$, in any direction $g \in R^m$, $|g|=1$.

The right-differentials $v_{h,g} \overset{\text{def}}{=} \delta_h^+ u(h,g)$ and $y_{h,g} \overset{\text{def}}{=} \delta_h^+ z(h,g)$ are given as a unique solution of the following quadratic optimal control problem

$$(QO_{h,g}) \quad \left| \begin{array}{l} \text{find } (v_{h,g}, y_{h,g}) \in \tilde{U} \times Z \text{ such that} \\[1ex] K(v_{h,g}, y_{h,g}; h,g) = \min\limits_{v \in \mathcal{V}_h^{ad}} K(v,y;h,g), \qquad (6.1.6) \\[1ex] \text{subject to} \\[1ex] y = S(h)v + (D_h S(h) g) u_h, \qquad (6.1.7) \end{array} \right.$$

where

$$K(v,y;h,g) = \frac{1}{2}\left[(v, \bar{Q}(h)v) + (y, \bar{\bar{Q}}(h)y)_Z\right] +$$

$$+ (\bar{q}(h,g), v) + (\bar{\bar{q}}(h,g), y)_Z \tag{6.1.8}$$

$$\bar{Q}(h) = D^2_{uu}F^1(u_h,h) + \sum_{i=1}^{r} \lambda^i_h D^2_{uu}\phi^i(u_h,h) = D^2_{uu}L(u_h,z_h;p_h,\lambda_h;h), \qquad (6.1.8a)$$

$$\bar{\bar{Q}}(h) = D^2_{zz}F^2(z_h,h) = D^2_{zz}L(u_h,z_h;p_h,\lambda_h;h), \qquad (6.1.8b)$$

$$\bar{q}(h,g) = D^2_{uh}F^1(u_h,h)g + \sum_{i=1}^{r} \lambda^i_h D^2_{uh}\phi^i(u_h,h)g + (D_hS^*(h)g)D_zF^2(z_h,h) =$$

$$= D^2_{uh}L(u_h,z_h;p_h,\lambda_h;h)g, \qquad (6.1.8c)$$

$$\bar{\bar{q}}(h,g) = D^2_{zh}F^2(z_h,h)g = D^2_{zh}L(u_h,z_h;p_h,\lambda_h;h)g, \qquad (6.1.8d)$$

$$\mathcal{V}^{ad}_h = \{v \in \tilde{U} \mid v(\xi) \in V^{ad}_h(\xi) \quad \text{for a.a. } \xi \in \Xi\}, \qquad (6.1.9)$$

$$\tilde{U} = \{v \in L^2(\Xi;R^n) \mid \bar{Q}(h)v \in L^2(\Xi;R^n)\}, \qquad (6.1.9a)$$

$$V^{ad}_h(\xi) = \{v \in R^n \mid \langle D_u\phi^i(u_h(\xi),h),v\rangle + \langle D_h\phi^i(u_h(\xi),h),g\rangle \begin{cases} =0 & \text{for } i \in I^c_h(\xi) \\ \leq 0 & \text{for } i \in I_h(\xi)\setminus I^c_h(\xi) \end{cases}\}$$

(6.1.9b)

The right-differentials of the Lagrange multipliers $r_{h,g} \overset{def}{=} \delta^+_h p(h,g)$ and $\mu_{h,g} \overset{def}{=} \delta^+_h \lambda(h,g)$ are given as the corresponding multipliers associated with $(QO_{h,g})$, as well as by the condition

$$\mu^i_{h,g}(\xi) = 0 \quad \text{for} \quad i \notin I_h(\xi). \qquad (6.1.10)$$

Lagrangian $L(u,z;p,\lambda;h)$ is defined here by (2.1.25).

Proof

By Theorem 2.1 we have

$$\|\tfrac{1}{\alpha}(u_{h+\alpha g}-u_h)\|, \|\tfrac{1}{\alpha}(z_{h+\alpha g}-z_h)\|_Y, \|\tfrac{1}{\alpha}(p_{h+\alpha g}-p_h)\|_Y, \|\tfrac{1}{\alpha}(\lambda_{h+\alpha g}-\lambda_h)\| \leq c\,.$$

(6.1.11)

Hence from any sequence $\{\alpha\}\downarrow 0$ we can extract a subsequence $\{\alpha'\} \subset \{\alpha\}$, such that for some element $r \in Y$

$$\frac{1}{\alpha'}(p_{h+\alpha' g}-p_h) \longrightarrow r$$

weakly in Y, and by (B1)

$$\frac{1}{\alpha'}(p_{h+\alpha' g}-p_h) \longrightarrow r \qquad (6.1.12)$$

strongly in Z.

Taking into account (B4), (B5) as well as (2.1.14) we find that

$$\frac{1}{\alpha'}\left[S^*(h+\alpha' g)p_{h+\alpha' g} - S^*(h)p_h\right] \longrightarrow S^*(h)r + (D_h S^*(h)g)p_h =$$

$$= S^*(h)r - (D_h S^*(h)g)D_z F^2(z_h,h)$$

strongly in $L^2(\Xi;R^n)$, which in particular implies the pointwise convergence

$$\frac{1}{\alpha'}\left[(S^*(h+\alpha' g)p_{h+\alpha' g})(\xi) - (S^*(h)p_h)(\xi)\right] \to (S^*(h)r)(\xi) +$$

$$- ((D_h S^*(h)g)D_z F^2(z_h,h))(\xi) \quad \text{for almost all } \xi \in \Xi. \tag{6.1.13}$$

Now let us return to the convex programming problem (CP_h) defined in Chapter 2, which pointwisely characterizes the solution u_h to (O_h).

Taking into account (B2), (B3), (B5), (B6) and (B9) as well as (6.1.13) and applying to (CP_h) exactly the same argument as in the proof of Theorem 4.1 we find that, for almost all $\xi \in \Xi$, there exist the limits

$$\lim_{\alpha' \downarrow 0} \frac{1}{\alpha'}\left[u_{h+\alpha' g}(\xi) - u_h(\xi)\right] = v(\xi), \tag{6.1.14a}$$

$$\lim_{\alpha' \downarrow 0} \frac{1}{\alpha'}\left[\lambda_{h+\alpha' g}(\xi) - \lambda_h(\xi)\right] = \mu(\xi), \tag{6.1.14b}$$

where $v(\xi)$ is a unique solution to the following quadratic programming problem

$$(QP_{h,g,\xi}) \quad \left| \begin{array}{l} \text{find } v(\xi) \in R^n \text{ such that} \\[1ex] k(v(\xi);h,g,\xi) = \min\limits_{v \in V_h^{ad}(\xi)} k(v;h,g,\xi), \end{array} \right. \tag{6.1.15}$$

where

$$k(v;h,g,\xi) = \frac{1}{2} < v, \bar{Q}(h,\xi)v > + < \rho(h,g,\xi), v >, \tag{6.1.16}$$

$$\bar{Q}(h,\xi) = D^2_{uu} f^1(u_h(\xi),h) + \sum_{i=1}^{r} \lambda_h^i D^2_{uu}\phi^i(u_h(\xi),h), \tag{6.1.16a}$$

$$\rho(h,g,\xi) = D^2_{uh} f^1(u_h(\xi),h) - (S^*(h)r)(\xi) + ((D_h S^*(h)g)D_z F^2(z_h,h))(\xi)$$

$$+ \sum_{i=1}^{r} \lambda_h^i D^2_{uh}\phi^i(u_h(\xi),h)g, \tag{6.1.16b}$$

$V_h^{ad}(\xi)$ - is defined by (6.1.9b).

For $i \in I_h(\xi)$ $\mu^i(\xi)$ are given as the unique Lagrange multipliers associated with $(QP_{h,g,\xi})$, while

$$\mu^i(\xi) = 0 \quad \text{for} \quad i \notin I_h(\xi). \tag{6.1.17}$$

Note that by the Lebesgue dominated convergence theorem [14] the pointwise convergence (6.1.14) and the estimates (6.1.11) imply

$$\frac{1}{\alpha'}(u_{h+\alpha' g} - u_h) \xrightarrow[\alpha' \downarrow 0]{} v \tag{6.1.18}$$

strongly in $L^2(\Xi;R^n)$,

$$\frac{1}{\alpha'}(\lambda_{h+\alpha' g} - \lambda_h) \xrightarrow[\alpha' \downarrow 0]{} \mu \tag{6.1.19}$$

strongly in $L^2(\Xi;R^r)$.

Taking advantage of (B5) we get from the state equation (2.1.3) and from (6.1.18)

$$\frac{1}{\alpha'}(z_{h+\alpha' g} - z_h) \xrightarrow[\alpha' \downarrow 0]{} y \tag{6.1.20}$$

strongly in Y, where

$$y = S(h)v + (D_h S(h)g)u_h. \tag{6.1.21}$$

Similarly from the adjoint equation (2.1.14) as well as from (6.1.20) and (B3) we obtain

$$\frac{1}{\alpha'}(p_{h+\alpha' g} - p_h) \xrightarrow[\alpha' \downarrow 0]{} r \tag{6.1.22}$$

strongly in Y, where

$$r = -D^2_{zz}F^2(z_h,h)y - D^2_{zh}F^2(z_h,h)g. \tag{6.1.23}$$

Of course, the limits (6.1.12) and (6.1.22) coincide.
It is easy to see that (6.1.15), (6.1.21) and (6.1.22) constitute necessary and sufficient conditions of optimality for $(QO_{h,g})$. Hence (v,y) is a solution to $(QO_{h,g})$, while (r,μ) are the associated Lagrange multipliers.

By (B2) Problem $(QO_{h,g})$ has a unique solution, while by (B9) and (6.1.10) the associated Lagrange multipliers are defined uniquely.

Therefore the limit elements in (6.1.18), (6.1.19), (6.1.20) and

(6.1.22) are independent of the choice of sequences $\{\alpha\}$ and $\{\alpha'\}$, hence they are equal to the respective right-differentials, i.e. (6.1.5) takes place and the theorem is proved.

□

Remark 6.1

It follows from (2.1.9) and (2.1.21) that the operator $\bar{Q}(h)$ defined by (6.1.8a) is continuous from $L^2(\Xi;R^n)$ into $L^1(\Xi;R^n)$, but it may not be continuous from $L^2(\Xi;R^n)$ into $L^2(\Xi;R^n)$. Therefore $(QO_{g,h})$ may not be well defined on $U \times Z$ and, instead of it, it is defined on $\tilde{U} \times Z$, where the subspace $\tilde{U} \subset U$ is given by (6.1.9a).

In a particular case, where control constraints are linear, the second term in the definition (6.1.8a) disappears and $\bar{Q}(h) \in \mathcal{L}(L^2(\Xi;R^n); L^2(\Xi;R^n))$. In this case $\tilde{U}=U$ in the formulation of $(QO_{h,g})$.

Note that the conditions of optimality for $(QO_{h,g})$, analogous to (2.1.14), (2.1.15), can be expressed in terms of the Lagrangian $L(u,z;p,\lambda;h)$ in the following simple form

$$D^2_{zz}L(u_h,z_h;p_h,\lambda_h;h)y_{h,g}+D^2_{zp}L(u_h,z_h;p_h,\lambda_h;h)r_{h,g}+D^2_{zh}L(u_h,z_h;p_h,\lambda_h;h)g=0 \tag{6.1.24a}$$

$$(D^2_{uu}L(u_h,z_h;p_h,\lambda_h;h)v_{h,g} + D^2_{up}L(u_h,z_h;p_h,\lambda_h;h)r_{h,g} +$$
$$+ D^2_{uh}L(u_h,z_h;p_h,\lambda_h;h)g, v-v_{h,g}) \geq 0 \qquad \forall v \in \mathcal{U}^{ad}_h. \tag{6.1.24b}$$

6.2. Abstract Problem. Continuous Differentiability

Using the results of Section 4.4 and repeating the proof of Theorem 6.1 we find that the left-differentials

$$v^-_{h,g} = \delta^-_h u(h,g) \quad \text{and} \quad y^-_{h,g} = \delta^-_h z(h,g)$$

of u_h and z_h at h in the direction g are given as the solution of the following quadratic optimal control problem

$$(QO^-_{h,g}) \quad \left| \begin{array}{l} \text{find } (v^-_{h,g}, y^-_{h,g}) \in \tilde{U} \times Z \text{ such that} \\[2ex] K(v_{h,g}, y_{h,g}; h,g) = \min\limits_{v \in \mathcal{U}^{ad-}_h} K(v,y;h,g), \\[2ex] \text{subject to} \\[1ex] y = S(h)v + D_h S(h)g)u_h \end{array} \right.$$

where $K(v,y;h,g)$ is given by (6.1.8), while $\mathcal{V}_h^{ad^-}$ is given by (6.1.9) with $V_h^{ad}(\xi)$ substituted by

$$V_h^{ad^-}(\xi)=\{v\in R^n \mid \langle D_u\phi^i(u_h(\xi),h),v\rangle+\langle D_h\phi^i(u_h(\xi),h),g\rangle \begin{cases} =0 & \text{for } i\in I_h^c(\xi) \\ \geq 0 & \text{for } i\in I_h(\xi)\setminus I_h^c(\xi) \end{cases}\}$$

Comparing $(QO_{h,g})$ and $(QO^-_{h,g})$ shows that in general $v_{h,g}\neq v^-_{h,g}$ and $y_{h,g}\neq y^-_{h,g}$. However in the case where

$$\text{meas } \{\xi\in\Xi \mid I_h(\xi)\setminus I_h^c(\xi)\neq\emptyset\}=0 \tag{6.2.1}$$

the solutions of $(QO_{h,g})$ and $(QO^-_{h,g})$ coincide for any direction $g\in R^m$. The same refers to the associated Lagrange multipliers.

Hence we obtain:

Proposition 6.1

If conditions (B1) through (B9) and (6.2.1) hold, then the solutions (u_h,y_h) to (O_h) and the associated Lagrange multipliers (p_h,λ_h) are continuously Gâteaux differentiable at h.

As in the case of convex programming (P_h) also for the Problems (O_h) the condition (6.2.1) is not needed to receiving the continuous differentiability of the optimal value function defined by

$$F^o(\cdot): H\longrightarrow R^1, \tag{6.2.2}$$

$$F^o(h) \overset{def}{=} F(u_h,z_h,h).$$

Indeed since by (2.1.3) and (2.1.20a) the second and the third terms in Lagrangian (2.1.25) vanish at $(u_h,z_h;p_h,\lambda_h;h)$, we obtain

$$F^o(h)=L(u_h,z_h;p_h,\lambda_h;h). \tag{6.2.3}$$

Hence

$$\delta^+_{h,g}F^o(h) = (D_uL(u_h,z_h;p_h,\lambda_h;h),v_{h,g})+(D_zL(u_h,z_h;p_h,\lambda_h;h),y_{h,g})+$$

$$+(D_pL(u_h,z_h;p_h,\lambda_h;h),r_{h,g})+(D_\lambda L(u_h,z_h;p_h,\lambda_h;h),\mu_{h,g})+$$

$$+\langle D_hL(u_h,z_h;p_h,\lambda_h;h),g\rangle. \tag{6.2.4}$$

Note that by (2.1.3) and (2.1.25) we get

$$(D_pL(u_h,z_h;p_h,\lambda_h;h),r_{h,g})=0. \tag{6.2.5}$$

Similarly by (6.1.4a) and (6.1.17)

$$(D_\lambda L(u_h, z_h; p_h, \lambda_h; h), \mu_{h,g}) = 0. \tag{6.2.6}$$

Substituting (2.1.27), (6.2.5) and (6.2.6) to (6.2.4) we obtain

$$\delta^+_{h,g} F^O(h) = \langle D_h L(u_h, z_h; p_h, \lambda_h; h), g \rangle . \tag{6.2.7a}$$

Similarly, for the left-derivative we get

$$\delta^-_{h,g} F^O(h) = \langle D_h L(u_h, z_h; p_h, \lambda_h; h), g \rangle . \tag{6.2.7b}$$

Note that by (B2), (B4) and (B6) as well as by Theorem 2.1

$$D_h L(u_h, z_h; p_h, \lambda_h, h) = D_h F^1(u_h, h) + D_h F^2(z_h, h) - (p_h, D_h S(h) u_h) + (\lambda_h, D_h \phi(u_h, h))$$

is a continuous function of h.

Hence from (6.2.7) we obtain

Proposition 6.2

If the conditions (B1) through (B9) hold, then the optimal value function $F^O(\cdot)$ for (P_h) is continuously (Fréchet) differentiable at any $h \in H$, and

$$D_h F^O(h) = D_h L(u_h, z_h; p_h, \lambda_h; h) \tag{6.2.8}$$

Note that the results of the type (6.2.8) are well known in stability analysis of optimal control problems (see e.g. [20, 42]).

6.3. Ordinary Differential Equations

In this section the abstract results of Section 6.1 will be used to the stability analysis of an optimal control problem, where the state equation is given by the system of ordinary differential equations

$$\dot{x}(t) = A(h)\, x(t) + B(h) u(t), \tag{6.3.1}$$

$$x(0) = 0,$$

where $x(t) \in R^\ell$, $u(t) \in R^n$.

The cost functional has the form

$$F(u,x,h) = F^1(u,h)+F^2(x,h) = \int_0^T f^1(u(t),h)\,dt + \int_0^T f^2(x(t),h)\,dt, \qquad (6.3.2)$$

where (0,T) is a fixed interval.
The control space $U = L^2(0,T;R^n)$.
We choose the spaces

$$Z = L^2(0,T;R^\ell) \quad \text{and} \quad Y = W^{1,2}(0,T;R^\ell),$$

where

$$W^{1,2}(0,T;R^\ell) = \{z \in L^2(0,T;R^\ell) \mid \dot{z} \in L^2(0,T;R^\ell)\} \qquad (6.3.3)$$

is the Sobolev space supplied with the norm

$$||z||_{1,2} = (||z||^2 + ||\dot{z}||^2)^{\frac{1}{2}} .$$

In terms of the abstract formulation of Chapter 2 the linear mapping

$$S(h) \in \mathcal{L}(L^2(0,T;R^n);\ W^{1,2}(0,T;R^\ell))$$

is given by the solution of the equation (6.3.1).

We consider the following problem of optimal control

$$(O_h')\qquad \left|\begin{array}{l} \text{find a pair } (u_h,x_h) \in L^2(0,T;R^n) \times L^2(0,T;R^\ell) \text{ such that} \\ F(u_h,x_h,h) = \min\limits_{u \in U_h^{ad}} F(u,x,h), \\ \text{subject to (6.3.1),} \end{array}\right.$$

where U_h^{ad} is given by (6.1.3) with $\Xi = [0,T]$.

We assume that the following conditions are satisfied:

(B'1) the conditions (B2) and (B3) hold,

(B'2) the matrix functions $A(\cdot)$ and $B(\cdot)$ are continuously differentiable on H,

(B'3) the conditions (B6)-(B9) hold.

Remarks 6.2

The condition (2.1.9) in the assumption (B8) can be droped as it is seen in Chapter 7, where the optimal control problem subject to control and state constraints is considered (see also [37]).

We are going to show that under the assumptions (B'1)-(B'3) all the assumptions (B1)-(B9) of Theorem 6.1 hold.

First let us note that by the Rellich-Kondrachov theorem [1] the embedding

$$W^{1,2}(0,T;R^{\ell}) \subset L^2(0,T;R^{\ell})$$

is compact . Hence it is enough to verify assumptions (B4) and (B5).

It is easy to see that for any $z \in Z = L^2(0,T;R^{\ell})$

$$S^*(h)z = B^T(h)p, \tag{6.3.4}$$

where p is the solution of the following equation adjoint to (6.3.1)

$$\dot{p}(t) = -A^T(h)p(t)-z(t). \tag{6.3.5}$$

$$p(T) = 0. \tag{6.3.5a}$$

Hence $S^*(h)$ is a continuous mapping from $L^2(0,T;R^{\ell})$ into $W^{1,2}(0,T;R^{\ell}) \subset L^4(0,T;R^{\ell})$ and the assumption (B4) holds.

Now one can check that for any direction $g \in R^m$

$$(D_h S(h)g)u = \rho, \tag{6.3.6}$$

where

$$\dot{\rho}(t) = A(h)\rho(t) + (D_h A(h)g)z(t) + (D_h B(h)g)u(t), \tag{6.3.7}$$

$$\rho(0) = 0, \tag{6.3.7a}$$

and the pair (u,z) satisfies (6.3.1).

Similarly from (6.3.4), (6.3.5) we obtain

$$(D_h S^*(h)g)z = (D_h B^T(h)g)p + B^T(h)\pi, \tag{6.3.8}$$

where

$$\dot{\pi}(t) = -A^T(h)\pi(t) - (D_h A^T(h)g)p(t), \tag{6.3.9}$$

$$\pi(T) = 0. \tag{6.3.9a}$$

It shows that the assumption (B5) holds.
Therefore we can apply Theorem 6.1 to find the form of the right-differentials of the solutions to (O_h').

To do that we have to find the form of all terms in (6.1.7) and (6.1.8).

Using (6.3.2) and (6.3.7) we find that the state equation (6.1.7) becomes

$$\dot{y}(t) = A(h)y(t)+B(h)v(t)+(D_hA(h)g)x_h(t)+(D_hB(h)g)u_h(t), \qquad (6.3.10)$$

$$y(0) = 0. \qquad (6.3.10a)$$

From (6.3.8) we get

$$-((D_hS^*(h)g)D_xF^2(x_h,h)v) = ((D_hB^T(h)g)p_h,v)+(\pi_h,B(h)v), \qquad (6.3.11)$$

where

$$\dot{p}_h = -A^T(h)p_h(t) + D_xf^2(x_h(t),h), \qquad (6.3.12)$$

$$p_h(T) = 0, \qquad (6.3.12a)$$

and π_h is given by (6.3.9), with $p(t)$ substituted by $p_h(t)$.

Taking advantage of (6.3.10), integrating by parts and using (6.3.9) we obtain

$$(\dot{\pi}_h,B(h)v) = (\pi_h,\dot{y}-A(h)y) - (\pi_h,(D_hA(h)g)x_h + (D_hB(h)g)u_h) =$$

$$= -(\dot{\pi}_h+A^T(h)\pi_h,y)-(\pi_h,(D_hA(h)g)x_h+(D_hB(h)g)u_h) =$$

$$= ((D_hA^T(h)g)p_h,y)-(\pi_h,(D_hA(h)g)x_h+(D_hB(h)g)u_h). \qquad (6.3.13)$$

Using (6.3.2), (6.3.11) and (6.3.13) we get from Theorem 6.1:

<u>Corollary 6.1</u>

If the assumptions (B'1)-(B'3) hold, then the solutions (u_h,x_h) of (O_h') and the associated multipliers (p_h,λ_h) are directionally differentiable functions of the parameter h, at any $h \in H$, in any direction $g \in R^m$, $|g|=1$.

The right-differentials $v_{h,g} \overset{\text{def}}{=} \delta_h^+u(h,g)$ and $y_{h,g} \overset{\text{def}}{=} \delta_h^+x(h,g)$ are given as a unique solution of the following quadratic optimal control problem

$$(QO'_{h,g}) \quad \left| \begin{array}{l} \text{find } (v_{h,g},y_{h,g}) \in U \times Z \text{ such that} \\ K(v_{h,g},y_{h,g}) = \min\limits_{v \in \mathcal{U}_h^{ad}} K(v,y;h,g), \\ \text{subject to (6.3.10),} \end{array} \right. \qquad (6.3.14)$$

where

$$K(v,y;h,g) = \int_0^T \{\tfrac{1}{2}[<(t),\bar{Q}_h(t)v(t)>+<y(t),\bar{\bar{Q}}_h(t)y(t)>] +$$
$$+ [<\bar{q}_{h,g}(t),v(t)>+<\bar{\bar{q}}_{h,g}(t),y(t)>]\}dt, \qquad (6.3.15)$$

$$\bar{Q}_h(t) = D^2_{uu}f^1(u_h(t),h) + \sum_{i=1}^{r} \lambda^i_h(t) D^2_{uu}\phi^i(u_h(t),h), \tag{6.3.15a}$$

$$\bar{\bar{Q}}_h(t) = D^2_{xx}f^2(x_h(t),h), \tag{6.3.15b}$$

$$\bar{q}_{h,g}(t) = D^2_{uh}f^1(u_h(t),h)g + \sum_{i=1}^{r} \lambda^i_h(t) D^2_{uh}\phi^i(u_h(t),h)g - (D_hB^T(h)g)p_h(t), \tag{6.3.15c}$$

$$\bar{\bar{q}}_{h,g}(t) = D^2_{xh}f^2(x_h(t),h)g - (D_hA^T(h)g)p_h(t), \tag{6.3.15d}$$

$$\mathcal{V}^{ad}_h = \{v \in L^2(0,T;R^n) \; v(t) \in V^{ad}_h(t)\}, \tag{6.3.16}$$

$$V^{ad}_h(t) = \{v \in R^n \mid \langle D_u\phi^i(u_h(t),h),v\rangle + \langle D_h\phi^i(u_h(t),h),g\rangle \begin{cases} =0 & \text{for } i \in I^c_h(t) \\ \leqslant 0 & \text{for } i \in I_h(t) \setminus I^c_h(t) \end{cases} \} \tag{6.3.16a}$$

p_h - is the solution to (6.3.12).

The right-differentials of the Lagrange multipliers $r_{h,g} \overset{\text{def}}{=} \delta^+_h p(h,g)$ and $\mu_{h,g} \overset{\text{def}}{=} \delta^+_h \lambda(h,g)$ are given as the corresponding multipliers associated with $(QO'_{h,g})$, as well as by

$$\mu^i_{h,g}(t) = 0 \quad \text{for} \quad i \notin I_h(t). \tag{6.3.17}$$

Remark 6.3

By Lemma 3.6 in our case the Lagrange multiplier λ_h is a uniformly bounded function on $(0,T)$, hence the operator $\bar{Q}(h)$ given by (6.1.8a) is continuous from $L^2(0,T;R^n)$ into $L^2(0,T;R^n)$. Therefore the subspace

$$\tilde{U} = \{u \in L^2(0,T;R^n) \mid \bar{Q}(h)u \in L^2(0,T;R^n)\}$$

coincides with the whole space $U = L^2(0,T;R^n)$.

In definition (6.3.15) of functional $K(v,y;h,g)$ the last term in (6.3.13) is ommited for it is independent of v and y and does not influence the solution to $(QO'_{h,g})$.

6.4. Boundary Control for Parabolic System

This section is devoted to the application of Theorem 6.1 to a boundary control problem for a system described by a partial differential equation of parabolic type.

In order to define the state equation, we have to introduce some functional spaces. Their precise definitions and properties can be found in [36].

Let $\Omega \subset R^2$ be a bounded open set, locally situated on one side of its boundary Γ, which is a smooth arc.

$[0,T]$ - is a fixed interval of control. Denote

$$\pi = \Omega \times (0,T), \quad \Sigma = \Gamma \times (0,T).$$

We define the following spaces:

$H^0(\Omega) = L^2(\Omega)$ - is the space of measurable functions, square integrable on Ω,

$H^\sigma(\Omega)$ - is the Sobolev spaces of fractional order σ defined on Ω,

$$\mathcal{H}^\sigma(\Omega) = \begin{cases} H^\sigma(\Omega) & \text{if} \quad \sigma \geq 0 \\ (H^{-\sigma}(\Omega))' & \text{if} \quad \sigma < 0, \end{cases}$$

where ' denotes the adjoint space.

$H^\rho(0,T;H^\sigma(\Omega))$ - is the Sobolev spaces of order ρ with respect to t having its range in $H^\sigma(\Omega)$,

$H^{\sigma,\rho}(\pi) = H^0(0,T;H^\sigma(\Omega)) \cap H^\rho(0,T;H^0(\Omega))$,

$$\mathcal{H}^{\sigma,\rho}(\pi) = \begin{cases} H^{\sigma,\rho}(\pi) & \text{if} \quad \sigma,\rho \geq 0, \\ (H^{-\sigma,-\rho}(\pi))' & \text{if} \quad \sigma,\rho < 0. \end{cases}$$

For sufficiently regular function z defined on π by

$$z|_\Sigma \quad \text{and} \quad z|_\Omega$$

we denote its traces on Σ and Ω respectively.

The spaces of functions defined on Σ are denoted in an analogous way as those on π.

We shall need the following results [36] concerning some properties of Sobolev spaces:

Lemma 6.1

If $z \in H^{2\rho,\rho}(\pi)$, then

$$z|_\Omega \in H^{2\rho-1}(\Omega) \qquad \text{for} \quad \rho > \frac{1}{2},$$

$$z|_\Sigma \in H^{2(\rho-\frac{1}{4}),\rho-\frac{1}{4}}(\Sigma) \qquad \text{for} \quad \rho > \frac{1}{4},$$

and the mappings

$$z \longrightarrow z|_{\Omega} \quad \text{and} \quad z \longrightarrow z|_{\Sigma}$$

are continuous in the respective topologies.

Lemma 6.2

If $\rho > 0$, then the embedding

$$H^{\rho}(\Omega) \subset L^2(\Omega)$$

is compact.

Let us consider the following Neumann-type boundary value problem for a parabolic equation defined on π:

$$D_t z(x,t) - Az(x,t) = f(x,t) \quad \text{in } \pi, \tag{6.4.1}$$

$$D_{\eta_A} z(x,t) = g(x,t) \quad \text{in } \Sigma, \tag{6.4.1a}$$

$$z(x,0) = z_o(x) \quad \text{in } \Omega, \tag{6.4.1b}$$

where

$$Az(x) = \sum_{i,j=1}^{2} D_{x_j}(a_{ij}(x) D_{x_i} y(x)) - a_o(x) y(x),$$

$$D_{\eta_A} z(x) = \sum_{i,j=1}^{2} a_{ij}(x) D_{x_j} y(x) \cos(\eta, x_i).$$

The functions $a_{ij}(\cdot) = a_{ji}(\cdot)$ are of class C^2 and satisfy the condition

$$\sum_{i,j=1}^{2} a_{ij}(x) \xi_i \xi_i \geqslant \delta(\xi_1^2 + \xi_2^2), \quad \delta > 0, \quad \forall x \in \Omega, \quad \forall \xi_1, \xi_2 \in R^1. \tag{6.4.2}$$

η is the unit outward vector normal to Γ.

The solution to (6.4.1) will be understood in the weak sens (see [36]) as the properly regular function satisfying the following identity

$$(D_t z(t), \psi) + a(z(t), \psi) = (f(t), \psi) + (g(t), \psi)_{\Gamma} \quad \forall \psi \in H^1(\Omega), \tag{6.4.3}$$

along with

$$z(0) = z_o, \tag{6.4.3a}$$

where

$$(z,\psi) \overset{\text{def}}{=} \int_{\Omega} y(x) \psi(x) dx, \quad (z,\psi)_{\Gamma} \overset{\text{def}}{=} \int_{\Gamma} z(x) \psi(x) dx,$$

$$a(z,\psi) \overset{\text{def}}{=} \int_{\Omega} \Big[\sum_{i,j=1}^{2} a_{ij}(x) D_{x_i} z(x) D_{x_j} \psi(x) + a_o(x) z(x) \psi(x) \Big] dx. \tag{6.4.4}$$

We shall need the following lemma concerning existence and requ-larity of the solutions to (6.4.3), which is a particular case of the results presented in [36]:

Lemma 6.3

Let $\rho \in (-1,0)$. If

$$f \in \mathcal{H}^{2\rho,\rho}(\pi),$$

$$g \in \mathcal{H}^{2(\rho+1/4),\rho+1/4}(\Sigma),$$

$$z_o \in \mathcal{H}^{2\rho+1}(\Omega),$$

then there exists a unique solution

$$z \in H^{2(\rho+1),\rho+1}(\pi)$$

of (6.4.3), which continuously depends on f,g and z_o.

Now we can formulate our optimal control problem.
For each $h \in H$ the state equation is given by

$$(D_t z(t),\psi)+a(z(t),\psi,h)=(u(t),\psi)_\Gamma \qquad \forall\psi \in H^1(\Omega), \tag{6.4.5}$$

$$z(0) = 0, \tag{6.4.5a}$$

where

$$a(z,\psi,h) \overset{def}{=} \int_\Omega \Big[\sum_{i,j=1}^{2} a_{ij}(x,h)D_{x_i}z(x)D_{x_j}\psi(x)+a_o(x,h)z(x)\psi(x)\Big]dx. \tag{6.4.6}$$

The cost functional has the form

$$F(u,z,h)=F^1(u,h)+F^2(z,h)=\int_\Sigma f^1(u(x,t),h)d\Sigma+\int_\Omega f^2(z(x,T),h)d\Omega. \tag{6.4.7}$$

The control space $U=L^2(\Sigma)$ and the set of admissible control is given by

$$U_h^{ad}=\{u \in L^2(\Sigma) \mid \phi^1(h) \leqslant u(\xi) \leqslant \phi^2(h) \quad \text{for a.a. } \xi=(x,t) \in \Sigma\}. \tag{6.4.8}$$

We choose the spaces

$$Z = L^2(\Omega), \quad Y = H^{1/2}(\Omega).$$

The linear mapping

$$S(h) : U \to Y$$

is given by the solution to (6.4.5).

For each $h \in H$ we consider the following problem of optimal control

$$(O_h'') \quad \left| \begin{array}{l} \text{find } (u_h, z_h) \in L^2(\Sigma) \times L^2(\Omega) \text{ such that} \\ F(u_h, z_h, h) = \min\limits_{u \in U_h^{ad}} F(u,z,h), \\ \text{subject to (6.4.5).} \end{array} \right.$$

We assume that the following conditions are satisfied:

(B"1) the conditions (B2) and (B3) hold,

(B"2) for each $h \in H$ the functions $a_{ij}(\cdot,h) = a_{ji}(\cdot,h)$ and $a_o(\cdot,h)$ are twice continuously differentiable and the condition (6.4.2) is satisfied uniformly with respect to h,

(B"3) the functions $a_{ij}(.,.)$, $a_o(.,.)$, $D_x a_{ij}(.,.)$, $D_x a_o(.,.)$ are continuously differentiable in all variables,

(B"4) the functions $\phi^i(h)$ are continuously differentiable and

$$\phi^1(h) < \phi^2(h) \qquad \forall h \in H.$$

We are going to show that (B"1)-(B"4) imply that all assumptions (B1)-(B9) of Theorem 6.1 hold.

First let us note that by Lemma 6.1 the solutions z of (6.4.5) satisfy

$$z \in H^{3/2,3/4}(\pi), \tag{6.4.9}$$

and by Lemma 6.1

$$z(T) \in H^{1/2}(\Omega).$$

Hence the mapping $S(h) : U \to Y$ is continuous

$$S(h) \in \mathcal{L}(L^2(\Sigma); H^{1/2}(\Omega)), \tag{6.4.10}$$

and by Lemma 6.1 the assumption (B1) holds.

Note that for any $z \in Z = L^2(\Omega)$

$$S^*(h)z = p|_\Sigma,$$

where p is the solution of the following equation adjoint to (6.4.5)

$$(D_t p(t), \psi) - a(p(t), \psi, h) = 0 \qquad \forall \psi \in H^1(\Omega), \tag{6.4.11}$$

$$p(T) = z. \tag{6.4.11a}$$

By Lemma 6.3

$$p \in H^{3/2,3/4}(\pi) \quad \text{if} \quad z \in H^{1/2}(\Omega), \tag{6.4.12a}$$

$$p \in H^{1,1/2}(\pi) \quad \text{if} \quad z \in L^2(\Omega), \tag{6.4.12b}$$

and by Lemma 6.1

$$S^*(h) \in \mathcal{L}(L^2(\Omega); H^{1/2,1/4}(\Sigma)) \cap \mathcal{L}(H^{1/2}(\Omega); H^{1,1/2}(\Sigma)).$$

Note that $H^{1/2,1/4}(\Sigma) \subset L^2(\Sigma)$,

while by a well known embedding theorem (see [52]), for $n=2$, we have

$$H^{1,1/2}(\Sigma) \subset L^4(\Sigma).$$

Therefore

$$S^*(h) \in \mathcal{L}(L^2(\Omega); L^2(\Sigma)) \cap \mathcal{L}(H^{1/2}(\Omega); L^4(\Sigma)). \tag{6.4.13}$$

By (6.4.10) and (6.4.13) the assumption (B4) is satisfied.

Now it is easy to see that for any direction $g \in R^m$

$$(D_h S(h) g) u = \zeta(T), \tag{6.4.14}$$

where ζ is the solution of the equation

$$(D_t \zeta(t), \psi) + a(\zeta(t), \psi, h) = -\langle D_h a(z(t), \psi, h), g \rangle, \quad \forall \psi \in H^1(\Omega) \tag{6.4.15a}$$

$$\zeta(0) = 0, \tag{6.4.15b}$$

and (z,u) satisfy (6.4.5), while

$$\langle D_h a(z(t), \psi, h), g \rangle \overset{\text{def}}{=} \int_\Omega \sum_{i,j=1}^{2} \langle D_h a_{ij}(x,h), g \rangle D_{x_i} z(x) D_{x_j} \psi(x) \, dx . \tag{6.4.16}$$

Similarly

$$(D_h S^*(h) g) z = \eta|_\Sigma , \tag{6.4.17}$$

where η is the solution of the equation

$$(D_t \eta(t), \psi) - a(\eta(t), \psi, h) = \langle D_h a(p(t), \psi, h), g \rangle \quad \forall \psi \in H^1(\Omega) \tag{6.4.18}$$

$$\eta(T) = 0, \tag{6.4.18a}$$

and p satisfies (6.4.11).

Note that if $z(t) \in H^\sigma(\Omega)$, then $\langle D_h a(z(t), \cdot, h), g \rangle$ is a linear continuous functional defined on $H^{2-\sigma}(\Omega)$.

Hence by (6.4.9) for almost all $t \in [0,T]$ the left hand side of (6.4.15) belongs to $\mathcal{H}^{-1/2}(\Omega)$ and by Lemma 6.3

$$\zeta \in H^{3/2,3/4}(\pi)$$

i.e. by Lemma 6.1 and by (6.4.14)

$$(D_h S(h)g)u \in H^{1/2}(\Omega). \tag{6.4.19}$$

Taking into account (6.4.12) and using the same argument as above we obtain

$$\eta \in H^{3/2,3/4}(\pi) \qquad \text{if} \qquad z \in H^{1/2}(\Omega),$$

$$\eta \in H^{1,1/2}(\pi) \qquad \text{if} \qquad z \in L^2(\Omega),$$

and by Lemma 6.1 together with (6.4.17)

$$(D_h S^*(h)g)z \in H^{1,1/2}(\Sigma) \subset L^4(\Sigma) \quad \text{if} \quad z \in H^{1/2}(\Omega), \tag{6.4.20a}$$

$$(D_h S^*(h)g)z \in H^{1/2,1/4}(\Sigma) \quad \text{if} \quad z \in L^2(\Omega). \tag{6.4.20b}$$

By (6.4.19) and (6.4.20) the assumption (B5) holds.

Finally it is obvious that (B"4) implies (B6)-(B9). Therefore all the assumptions (B1) through (B9) are satisfied and we can apply Theorem 6.1 to (O''_h).

To this end we must find the form of all terms in (6.1.7)-(6.1.9). Using (6.4.5) and (6.4.15) we find that the state equation (6.1.7) is given by

$$(D_t y(t),\psi)+a(y(t),\psi,h)=-\langle D_h a(z_h(t),\psi,h)g\rangle+(v(t),\psi)_\Gamma \qquad \forall\psi \in H^1(\Omega), \tag{6.4.21}$$

$$y(0) = 0. \tag{6.4.21a}$$

From (6.4.17) we get

$$((D_h S^*(h)g)D_z F^2(z_h,h),v)_\Sigma = (\eta_h,v)_\Sigma, \tag{6.4.22}$$

where η_h is the solution of (6.4.18) with p substituted by p_h, which satisfies

$$(D_t p_h(t),\psi)-a(p_h(t),\psi,h) = 0 \qquad \forall\psi \in H^1(\Omega) \tag{6.4.23}$$

$$p_h(T) = D_z F^2(z_h,h). \tag{6.4.23a}$$

Let us substitute (6.4.21) into (6.4.22). Integrating by parts with respect to t, taking advantage of the symmetry of $a(.,.,h)$ and using (6.4.18) we get

$$((D_h S^*(h)g)D_z F^2(z_h,h),v) = (\eta_h,v)_\Sigma = \int_0^T (D_t y(t),\eta_h(t))dt +$$

$$+\int_0^T a(y(t),\eta_h(t),h)dt + \int_0^T < D_h a(z_h(t),\eta_h(t),h),g > dt =$$

$$=-\int_0^T < D_h a(p_h(t),y(t),h),g > dt + \int_0^T < D_h a(z_h(t),\eta_h(t),h),g > dt. \quad (6.4.24)$$

Finally from (6.4.8) it follows that the set (6.1.9b) takes on the form

$$V_h^{ad}(\xi)=\{v\in R^1 \,|\, v=D_h\phi^i(h) \quad \text{if} \quad u_h(\xi)=\phi^i(h),\ \lambda_h^i(\xi)>0,\ i=1,2,$$
$$v\geqslant D_h\phi^1(h) \quad \text{if} \quad u_h(\xi)=\phi^1(h),\ \lambda_h^1(\xi)=0,$$
$$v\leqslant D_h\phi^2(h) \quad \text{if} \quad u_h(\xi)=\phi^2(h),\ \lambda_h^2(\xi)=0\}, \quad (6.4.25)$$

where $\lambda_h^i\in L^2(\Sigma)\,(i=1,2)$ are Lagrange multipliers associated with the lower and the upper boundes in (6.4.8).

Using (6.4.7), (6.4.24) and (6.4.25) we obtain from Theorem 6.1:

Corollary 6.2

If the assumptions (B"1) through (B"4) hold, then the solutions (u_h,z_h) of (O_h'') and the associated Lagrange multipliers p_h,λ_h are directionally differentiable functions of the parameter h, at any $h\in H$, in any direction $g\in R^m$, $|g|=1$.

The right-differentials $v_{h,g}=\delta_h^+u(h,g)$ and $y_{h,g}=\delta_h^+z(h,g)$ are given as a unique solution of the following quadratic optimal control problem

$(QO''_{h,g})$ find $(v_{h,g},y_{h,g})\in L^2(\Sigma)\times L^2(\pi)$

such that

$$K(v_{h,g},y_{h,g};h,g) = \min_{v\in \mathcal{V}_h^{ad}} K(v,y;h,g), \quad (6.4.26)$$

subject to (6.4.21),

where

$$K(v,y;h,g) = \int_\Sigma \left[\tfrac{1}{2}\bar{Q}_h(\xi)v^2(\xi)+\bar{q}_h(\xi)v(\xi)\right]d\xi +$$

$$+\int_\Omega\left[\tfrac{1}{2}\bar{\bar{Q}}_h(x)y^2(x,T)+\bar{\bar{q}}_h(x)y(x,T)\right]dx - \int_0^T < D_h a(p_h(t),y(t),h),g > dt, \quad (6.4.27)$$

$$\bar{Q}_h(\xi) = D_{uu}^2 f^1(u_h(\xi),h), \quad (6.4.27a)$$

$$\bar{\bar{Q}}_h(x) = D^2_{zz}f^2(z_h(x,T),h), \tag{6.4.27b}$$

$$\bar{q}_{h,g}(\xi) = D^2_{uh}f^1(u_h(\xi),h)g, \tag{6.4.27c}$$

$$\bar{\bar{q}}_{h,g}(x) = D^2_{zh}f^2(z_h(x,T)),h)g\ , \tag{6.4.27d}$$

$$\mathcal{V}^{ad}_h=\{v\in L^2(\Xi)\ |\ v(\xi)\in V^{ad}_h(\xi) \quad \text{for a.a.} \quad \xi\in\Sigma\}, \tag{6.4.28}$$

with $V^{ad}_h(\xi)$ given by (6.4.25).
p_h - is the solution to (6.4.23).
The right-differentials of the Lagrange multipliers $r_{h,g} \overset{\text{def}}{=} \delta^+_h p(h,g)$ and $\mu_{h,g} \overset{\text{def}}{=} \delta^+_h \lambda(h,g)$ are given as the corresponding multipliers associated with $(QO''_{h,g})$ as well as by

$$\mu^i_{h,g}(\xi)=0 \quad \text{if} \quad \phi^1(h) < u_h(\xi) < \phi^2(h). \tag{6.4.29}$$

Note that since in our case the control constraints are linear, then according to Remark 6.1 we have

$$\tilde{U} = U = L^2(\Sigma).$$

In the definition (6.4.27) of the functional $K(v,y;h,g)$ the second term in (6.4.24) is omitted for it is independent of v and y and does not influence the solution to $(QO''_{h,g})$.

7. DIFFERENTIAL STABILITY OF SOLUTIONS TO OPTIMAL CONTROL PROBLEMS SUBJECT TO STATE AND CONTROL CONSTRAINTS

7.1. Right-Differentiability of Solutions

Let us recall the family $\{OC_h\}$ of the convex optimal control problems (OC_h), depending on a vector parameter $h \in H \subset R^m$, which was defined in Chapter 3;

(OC_h)

find a pair $(u_h, x_h) \in L^2(0,T) \times W^{1,1}(0,T)$ such that

$$F(u_h, x_h, h) = \min\{F(u,x,h) \overset{def}{=} \int_0^T f(u(t), x(t), h)\, dt\}, \tag{7.1.1}$$

subject to

$$\dot{x}(t) = A(h)\, x(t) + B(h) u(t), \tag{7.1.2}$$

$$x(0) = x^o, \tag{7.1.2a}$$

and

$$u(t) \in \Phi_h \overset{def}{=} \{u \in R^n \mid \phi^i(u,h) \leqslant 0,\ i \in I\} \quad \text{for a.a. } t \in [0,T], \tag{7.1.3}$$

$$x(t) \in \Theta_h \overset{def}{=} \{x \in R^\ell \mid \theta^j(x,h) \leqslant 0,\ j \in J\} \quad \text{for all } t \in [0,T]. \tag{7.1.4}$$

We assume that the conditions (C1) through (C8) hold, hence (OC_h) has a unique solution. To characterize this solution we shall use the Lagrangian L_2 defined by (3.2.26). By Corollary 3.9 the Lagrange multipliers $p_h, \lambda_h, \pi_h, \sigma_h$ associated with (OC_h) are defined uniquely.

For each $i \in I$ we define the sets

$$\Sigma^i(h) = \{t \in [0,T] \mid \phi^i(u_h(t),h) = 0\}, \tag{7.1.5a}$$

$$\Sigma^i_c(h) = \{t \in \Sigma^i(h) \mid \lambda^i_h(t) > 0\}. \tag{7.1.5b}$$

Similarly for each $j \in J$ we define

$$\Xi^j(h) = \{t \in [0,T] \mid \theta^j(x_h(t),h) = 0\}, \tag{7.1.6a}$$

$$\Xi^j_c(h) = \{t \in \Xi^j(h) \mid \pi^j_h(\cdot) \text{ - is increasing in a neigbourhood of } t\}. \tag{7.1.6b}$$

In order to be able to obtain results concerning differentiability of solutions to (OC_h) we have to add to (C1)-(C8) an additional restrictive assumption:

(C9) at the point $h \in H$

$$\text{meas}(\Sigma^i(h) \setminus \Sigma_c^i(h)) = 0 \qquad \forall i \in I.$$

It is obvious that the assumption (C9) becomes irrelevant in the case where control constraints are not present.

Our purpose is to show that the solutions (u_h, x_h) to (OC_h) and the associated Lagrange multipliers $p_h, \lambda_h, \pi_h, \sigma_h$ are right-differentiable functions of $h \in H$. To be more precise we are going to prove that under the assumptions (C1)-(C9), for any direction $g \in R^m$, $|g|=1$, the following limits exist

$$\delta_h^+ u(h,g) = w - \lim_{\alpha \downarrow 0} \frac{1}{\alpha}(u_{h+\alpha g} - u_h), \tag{7.1.7a}$$

$$\delta_h^+ x(h,g) = \lim_{\alpha \downarrow 0} \frac{1}{\alpha}(x_{h+\alpha g} - x_h), \tag{7.1.7b}$$

$$\delta_h^+ p(h,g) = w - \lim_{\alpha \downarrow 0} \frac{1}{\alpha}(p_{h+\alpha g} - p_h), \tag{7.1.7c}$$

$$\delta_h^+ \lambda(h,g) = w - \lim_{\alpha \downarrow 0} \frac{1}{\alpha}(\lambda_{h+\alpha g} - \lambda_h), \tag{7.1.7d}$$

$$\delta_h^+ \pi(h,g) = w - \lim_{\alpha \downarrow 0} \frac{1}{\alpha}(\pi_{h+\alpha g} - \pi_h), \tag{7.1.7e}$$

$$\delta_h^+ \sigma(h,g) = \lim_{\alpha \downarrow 0} \frac{1}{\alpha}(\sigma_{h+\alpha g} - \sigma_h), \tag{7.1.7f}$$

where in (7.1.7a), (7.1.7c), (7.1.7d) and (7.1.7e) the limits are taken in the weak topology of the space $L^2(0,T)$, while in (7.1.7b) the strong convergence in $C(0,T)$ and in (7.1.7f) the usual convergence in R^s take place.

The following theorem shows that the limits (7.1.7) exist and are characterized as the solutions and the associated Lagrange multipliers for an auxiliary quadratic optimal control problem subject to linear state and control constraints:

<u>Theorem 7.1</u>

If the assumptions (C1) through (C9) hold, then the solutions (u_h, x_h) of (OC_h) and the associated Lagrange multipliers $p_h, \lambda_h, \pi_h, \sigma_h$ are directionally differentiable functions of the parameter h, at $h \in H$, in any direction $g \in R^m$, $|g|=1$. Differentiability is understood in the sense of (7.1.7).

The right-differentials $v_{h,g} \overset{\text{def}}{=} \delta_h^+ u(h,g)$ and $y_{h,g} \overset{\text{def}}{=} \delta_h^+ x(h,g)$ are given as a unique solution of the following quadratic optimal control problem:

$$
(QC_{h,g})\quad
\begin{cases}
\text{find } (v_{h,g},y_{h,g}) \in L^2(0,T) \times W^{1,1}(0,T) \quad \text{such that} \\
K(v_{h,g},y_{h,g};h,g) = \min K(v,y;h,g), & (7.1.8)\\
\text{subject to} \\
\dot{y}(t) = A(h)y(t) + B(h)v(t) + (D_hA(h)g)x_h(t) + (D_hB(h)g)u_h(t), & (7.1.9)\\
y(0) = 0, & (7.1.9a)\\
\text{and} \\
v(t) \in V^{ad}_{h,g}(t) \quad \text{for almost all} \quad t \in [0,T], & (7.1.10)\\
y(t) \in Y^{ad}_{h,g}(t) \quad \text{for all} \quad t \in [0,T], & (7.1.11)
\end{cases}
$$

where

$$
K(v_{h,g},y_{h,g};h,g) = \int_0^T \{ \tfrac{1}{2}[v^T(t),y^T(t)] \begin{bmatrix} Q^{11}_h(t), Q^{12}_h(t) \\ Q^{21}_h(t), Q^{22}_h(t) \end{bmatrix} \begin{bmatrix} v(t) \\ y(t) \end{bmatrix} + \langle \bar{q}_{h,g}(t), v(t) \rangle +
$$

$$
+ \langle \bar{\bar{q}}_{h,g}(t), y(t) \rangle \} dt + \tfrac{1}{2} \langle y(T), S_h y(T) \rangle + \langle s_{h,g}, y(T) \rangle , \tag{7.1.12}
$$

$$
Q^{11}_h(t) = D^2_{uu} f(u_h(t),x_h(t),h) + \sum_{i=1}^{r} \lambda^i_h(t) D^2_{uu}\phi^i(u_h(t),h), \tag{7.1.12a}
$$

$$
Q^{12}_h(t) = [Q^{21}_h(t)]^T = D^2_{ux} f(u_h(t),x_h(t),h), \tag{7.1.12b}
$$

$$
Q^{22}_h(t) = D^2_{xx} f(u_h(t),x_h(t),h) + \sum_{j=1}^{s} \dot{\pi}^j(t) D^2_{xx}\theta^j(x_h(t),h), \tag{7.1.12c}
$$

$$
\bar{q}_{h,g}(t) = D^2_{uh} f(u_h(t),x_h(t),h)g + \sum_{i=1}^{r} \lambda^i_h(t) D^2_{uh}\phi^i(u_h(t),h))g - (D_hB^T(h)g)p_h(t), \tag{7.1.12d}
$$

$$
\bar{\bar{q}}_{h,g}(t) = D^2_{xh} f(u_h(t),h)g + \sum_{j=1}^{s} \dot{\pi}^j_h(t) D^2_{xh}\theta^j(x_h(t),h))g - (D_hA^T(h)g)p_h(t), \tag{7.1.12e}
$$

$$
S_h = \sum_{j=1}^{r} (\sigma^j_h - \pi^j_h(T)) D^2_{xx}\theta^j(x_h(T),h), \tag{7.1.12f}
$$

$$
s_h = \sum_{j=1}^{s} (\sigma^j_h - \pi^j_h(T)) D^2_{xh}\theta^j(x_h(T),h)g, \tag{7.1.12g}
$$

$$
V^{ad}_{h,g}(t) = \{ v \in R^n \mid \langle D_u\phi^i(u_h(t),h), v \rangle + \langle D_h\phi^i(u_h(t),h), g \rangle = 0 \ \text{ if } \ t \in \Sigma^i(h) \}, \tag{7.1.13}
$$

$$
Y^{ad}_{h,g}(t) = \{ y \in R^\ell \mid \langle D_x\theta^j(x_h(t),h), y \rangle + \langle D_h\theta^j(x_h(t),h), g \rangle \begin{cases} = 0 & \text{if } t \in \Xi^j_c(h) \\ \leq 0 & \text{if } t \in \Xi^j(h) \setminus \Xi^j_c(h) \end{cases} \}.
$$

(7.1.14a)
(7.1.14b)

The right-differentials of the Lagrange multipliers $r_{h,g} \overset{def}{\equiv} \delta_h^+ p(h,g)$, $\mu_{h,g} \overset{def}{\equiv} \delta_h^+ \lambda(h,g)$, $\rho_{h,g} \overset{def}{\equiv} \delta_h^+ \pi(h,g)$, $\zeta_{h,g} \overset{def}{\equiv} \delta_h^+ \rho(h,g)$ are given as the corresponding multipliers associated with $(QC_{h,g})$ together with the additional conditions

$$\mu^i_{h,g}(t) = 0 \quad \text{for almost all} \quad t \notin \Sigma^i_c(h), \tag{7.1.15a}$$

$$\rho^j_{h,g} \text{ - is constant in a neighbourhood of t if } \quad t \notin \Xi^j(h). \tag{7.1.15b}$$

Remark 7.1

The uniqueness of the solution to $(QC_{h,g})$ (if such a solution exists) follows from (C1), (C4), and the fact that π_h is a non-decreasing, Lipschitz continuous function. On the other hand the uniquness of the Lagrange multipliers can be obtained using (C8) in a way similar to the proof of Theorem 3.2.

Proof of Theorem 7.1

The proof is very similar to that given in [38], where a linear-quadratic problem subject only to state space constraints is considered. Let $h \in H$ be any arbitrary value of the parameter and let $g \in R^m$, $|g| = 1$ be any direction. By Theorem 3.2 and Corollary 3.7 for any arbitrary sequence $\{\alpha\} \downarrow 0$, we have

$$||\tfrac{1}{\alpha}(u_{h+\alpha g} - u_h)||, ||\tfrac{1}{\alpha}(\dot{x}_{h+\alpha g} - \dot{x}_h)||, ||\tfrac{1}{\alpha}(x_{h+\alpha g} - x_h)||_\infty, ||\tfrac{1}{\alpha}(p_{h+\alpha g} - p_h)||,$$

$$||\tfrac{1}{\alpha}(\lambda_{h+\alpha g} - \lambda_h)||, ||\tfrac{1}{\alpha}(\pi_{h+\alpha g} - \pi_h)||, |\tfrac{1}{\alpha}(\sigma_{h+\alpha g} - \sigma_h)| \leqslant c. \tag{7.1.16}$$

Hence by weak compactness of a closed ball in a Hilbert space and by the Ascoli-Arzela theorem [14] from any sequence $\{\alpha\}$ we can extract a subsequence $\{\alpha'\} \subset \{\alpha\}$, such that

$$\frac{1}{\alpha'}(u_{h+\alpha' g} - u_h) \xrightarrow[\alpha' \downarrow 0]{} v, \tag{7.1.17a}$$

$$\frac{1}{\alpha'}(\dot{x}_{h+\alpha' g} - \dot{x}_h) \xrightarrow[\alpha' \downarrow 0]{} \dot{y}, \tag{7.1.17b}$$

$$\frac{1}{\alpha'}(p_{h+\alpha' g} - p_h) \xrightarrow[\alpha' \downarrow 0]{} r, \tag{7.1.17c}$$

$$\frac{1}{\alpha'}(\lambda_{h+\alpha' g} - \lambda_h) \xrightarrow[\alpha' \downarrow 0]{} \mu, \tag{7.1.17d}$$

$$\frac{1}{\alpha'}(\pi_{h+\alpha' g}-\pi_h) \longrightarrow \rho \tag{7.1.17e}$$

weakly in $L^2(0,T)$, while

$$\frac{1}{\alpha'}(x_{h+\alpha' g}-x_h) \longrightarrow y \tag{7.1.17f}$$

strongly in $C(0,T)$, where y is a Lipschitz continuous function,

$$\frac{1}{\alpha'}(\sigma_{h+\alpha' g}-\sigma_h) \longrightarrow \zeta \tag{7.1.17g}$$

in R^s.

We are going to prove that the limit elements in (7.1.17) satisfy necessary and sufficient conditions of optimality for $(QC_{h,g})$, in the form analogous to (3.2.28)-(3.2.31).

Since the solution and the associated Lagrange multipliers for $(QC_{h,g})$ are unique, then the limit elements are independent of the choice of sequences $\{\alpha\}$ and $\{\alpha'\}$, therefore they are equal to the respective right-differentials (7.1.7) and the theorem is proved.

First of all we shall show that constraints (7.1.9)-(7.1.11) hold.

Indeed taking the difference quotient of (7.1.2) at $(h+\alpha' g)$ and at h, letting $\alpha'\downarrow 0$ and using (C3) and (7.1.17) we arrive at (7.1.9).

Let us prove (7.1.10).

(1) For a fixed $\gamma > 0$ let us denote

$$M^i_\gamma = \{t \in \Sigma^i_c(h) \mid \lambda^i_h(t) > \gamma\}.$$

and suppose that meas $M^i_\gamma > 0$.
By Theorem 3.2 for $\alpha' > 0$ sufficiently small we have

$$\lambda^i_{h+\alpha' g}(t) > \frac{\gamma}{2} \qquad \forall t \in M^i_\gamma.$$

Hence by the complementary slackness (3.2.30) we have

$$\phi^i(u_{h+\alpha' g}(t),h) = 0 \quad \text{for almost all} \quad t \in M^i_\gamma,$$

and by (C5) together with (7.1.16) and (7.1.17a) we find that (7.1.13) holds on M_γ. Since $\gamma > 0$ is an arbitrary number we find that (7.1.13) is satisfied on Σ^i_c. Finally by (C9) we obtain (7.1.13).

To prove (7.1.11) we consider two cases corresponding to (7.1.14a) and (7.1.14b).

(2) Let $t \in \Xi^j(h)$, i.e. $\theta^j(x_h(t),h)=0$. On the other hand for any

$(h+\alpha'g)\in H$ we have $\theta^j(x_{h+\alpha'g}(t),h\leq 0$, hence by (C5), (7.1.16) and (7.1.17) we find that (7.1.14b) holds on $\Xi^j(h)$.

(3) Let $t\in \Xi^j_c(h)\setminus \Xi^j(h)$, then by (7.1.14b) we have

$\langle D_x\theta^j(x_h(t),h),y(t)\rangle+\langle D_h\theta^j(x_h(t),h),g\rangle = \gamma\leq 0$. Suppose that $\gamma< 0$.

Since the set $\Xi^j_c(h)$ is open and the functions x_h and y are continuous, then there exists an open interval $M\subset \Xi^j_c(h)$ such that $t\in M$ and $\langle D_x\theta^j(x_h(\tau),h),y(\tau)\rangle+\langle D_x\theta^j(x_h(\tau),h),g\rangle\leq\frac{\gamma}{2}$ for any $\tau\in M$, which together with (C5), (7.1.16) and (7.1.17) imply that

$\frac{1}{\alpha'}[\theta^j(x_{h+\alpha'g}(\tau),h+\alpha'g)-\theta^j(x_h(\tau),h)]\leq\frac{\gamma}{4}$ for $\alpha'>0$ sufficiently small. Hence $\theta^j(x_{h+\alpha'g}(\tau),h+\alpha'g)\leq\frac{\gamma\alpha'}{4}$ for any $\tau\in M$ and by (3.2.35a) $\pi^j_{h+\alpha'g}(\tau)=\text{const}$ on M for $\alpha'>0$ sufficiently small. On the other hand by the definition (7.1.16b) $\pi^j_h(\tau)$ is increasing on M. These two facts violate (7.1.16). The obtained contradiction shows that (7.1.14a) holds on $\Xi^j_c(h)$. Since $x_h(\cdot)$ and $y(\cdot)$ are continuous functions this equality is satisfied on $\overline{\Xi}^j_c(h)$.

Now let us take the difference quotients of (3.2.28) and (3.2.29) at $(h+\alpha'g)$ and at h. Letting $\alpha'\downarrow 0$ and using (C2), (C3), (C5), (7.1.16) and (7.1.17) we obtain

$$\begin{aligned}
& r(t)-\int_t^T[A^T(h)r(\tau)+(D_hA^T(h)g)p_h(\tau)-D^2_{xu}f(u_h(\tau),x_h(\tau),h)v(\tau)+\\
& -D^2_{xx}f(u_h(\tau),x_h(\tau),h)y-D^2_{xh}f(u_h(\tau),x_h(\tau),h)g+\frac{d}{d\tau}(D^2_{xx}\theta^T(x_h(\tau),h)y(\tau)+\\
& +(D^2_{xh}\theta^T(x_h(\tau),h)g)\pi_h(\tau)+\frac{d}{d\tau}(D_x\theta^T(x_h(\tau),h))\rho(\tau)]d\tau+\\
& -(D^2_{xx}\theta^T(x_h(t),h)y(t)+(D^2_{xh}\theta^T(x_h(t)h))g)\pi_h(t)-D_x\theta^T(x_h(t),h)\rho(t)+\\
& +(D^2_{xx}\theta^T(x_h(T),h)y(T)+D^2_{xh}\theta^T(x_h(T),h)g)\sigma_h+D_x\theta^T(x_h(T),h)\zeta=\\
& =r(t)-\int_t^T[A^T(h)r(\tau)-D^2_{xu}f(u_h(\tau),x_h(\tau),h)v(\tau)+\\
& -(D^2_{xx}f(u_h(\tau),x_h(\tau),h)+D^2_{xx}\theta^T(x_h(\tau),h)\dot{\pi}_h(\tau))y(\tau)+(D_hA^T(h)g)p_h(\tau)+\\
& -(D^2_{xh}f(u_h(\tau),x_h(\tau),h)+D^2_{xh}\theta^T(x_h(\tau),h)\dot{\pi}_h(\tau))g+\frac{d}{d\tau}(D_x\theta^T(x_h(\tau),h))\rho(\tau)]d\tau+\\
& +D^2_{xx}\theta^T(x_h(T),h)(\sigma_h-\pi_h(T))y(T)+(D^2_{xh}\theta^T(x_h(T),h)g)(\sigma_h-\pi_h(T))+\\
& -D_x\theta^T(x_h(t),h)\rho(t)+D_x\theta^T(x_h(T),h)\zeta=0 \quad \text{for almost all } t\in[0,T], \quad (7.1.18)
\end{aligned}$$

$$(D^2_{uu}f(u_h(t),x_h(\tau),h)+D^2_{uu}\phi^T(u_h(t),h)\lambda_h(t))v(t)+D^2_{ux}f(u_h(t),x_h(t),h)y(t)+$$

$$+(D^2_{uh}f(u_h(t),x_h(t),h)+D^2_{uh}\phi^T(u_h(t),h)\lambda_h(t))g-(D_hB^T(h)g)p_h(t)+$$

$$-B^T(h)r(t)+D_u\phi^T(u_h(t),h)\mu(t)=0 \quad \text{for almost all} \quad t\in[0,T]. \tag{7.1.19}$$

It is easy to see that (7.1.18) and (7.1.19) constitute necessary and sufficient conditions of optimality for $(QC_{h,g})$ provided that μ, ρ and ζ satisfy the conditions of respective Lagrange multipliers and that the complementary slackness holds.

Namely μ^i must satisfy the condition

$$\mu^i(t)=0 \quad \text{for almost all} \quad t\in(0,T)\setminus\Sigma^i(h), \tag{7.1.20}$$

while ρ^j and σ^j must satisfy on $[0,T]\setminus\Xi^j_c(h)$ the conditions analogous to (3.2.32), (3.2.34):

ρ^j - is non-decreasing a.e. in a neighbourhood of any $t\in(0,T)\setminus\Xi^j_c(h)$, (7.1.21a)

$$\operatorname*{ess\,inf}_{t\in(0,\tau^j)}\rho^j(t)=0, \tag{7.1.21b}$$

$$\zeta^j\geqslant\operatorname*{ess\,sup}_{t\in(\xi^j,T)}\rho^j(t) \quad \text{if} \quad T\in\Xi^j_c(h) \quad \text{and} \quad \sigma^j_h=\operatorname*{ess\,sup}_{t\in[0,T]}\pi^j(t), \tag{7.1.21c}$$

where $0<\tau^j<\xi^j<T$ and

$$(0,\tau^j)\subset(0,T)\setminus\Xi^j_c, \quad (\xi^j,T)\subset(0,T)\setminus\Xi^j_c. \tag{7.1.21d}$$

Moreover the following complementary slackness condition analogous to (3.2.31) must hold:

$$-(\rho,\frac{d}{dt}(D_x\theta(x_h,h)y+D_h\theta(x_h,h)g))+\langle\zeta,D_x\theta(x_h(T),h)y(T)+D_h\theta(x_h(T),h)g\rangle=0. \tag{7.1.22}$$

Let us start with proving (7.1.20).

(4) For any $\gamma>0$ let us denote

$$M^i_\gamma=\{t\in(0,T)\setminus\Sigma^i(h)\,|\,\phi^i(u_h(t),h)<-\gamma\}.$$

By Theorem 3.2 for $\alpha'>0$ sufficiently small we have

$$\phi^i(u_{h+\alpha' g}(t))\leqslant-\frac{\gamma}{2} \qquad \forall t\in M^i_\gamma$$

and by the complementary slackness (3.2.30) we get

$$\lambda_h^i(t) = \lambda_{h+\alpha' g}^i(t) = 0 \qquad \forall t \in M_\gamma^i$$

for $\alpha' > 0$ sufficiently small, which together with (7.1.17d) show that $\mu^i(t)=0$ on M_γ^i. Since $\gamma > 0$ is arbitrary we get (7.1.20).

Now let us prove (7.1.21).

(5) Since the set $(0,T)\setminus \Xi_c^j$ is open, then for any $t \in (0,T)\setminus \Xi_c^j$ there exists a subinterval $M \subset (0,T)\setminus \Xi_c^j$ such that $t \in M$ and $\pi_h(\tau)=\text{const}$ for all $\tau \in M$. On the other hand $\pi_{h+\alpha' g}$ is non-decreasing on $[0,T]$. Hence by (7.1.17e) we get (7.1.21a).

To prove (7.1.21b) note that by (C5) and (C6) for any compact set $\mathcal{H} \subset H$ there exists $\tau^j > 0$ such that

$$\theta^j(x_g(t),g) < 0 \qquad \forall t \in [0,\tau^j], \qquad \forall g \in \mathcal{H},$$

hence by (3.2.34) and (3.2.35a)

$$\pi_g^j(t) = 0 \qquad \forall t \in [0,\tau^j], \qquad \forall g \in \mathcal{H},$$

which together with (7.1.17e) imply (7.1.21b).

Finally if $T \notin \Xi_c^j(h)$, then by (3.2.32b), (7.1.6b) and by Corollary 3.5, for any $\xi^j < T$ such that $[\xi^j,T) \subset (0,T)\setminus \Xi_c^j(h)$, we have

$$\pi_h^j(t) = \pi_h^j(T) \qquad \forall t \in [\xi^j,T]. \tag{7.1.23}$$

On the other hand by (3.2.32b)

$$\operatorname*{ess\,sup}_{t \in [0,T]} \pi_{h+\alpha' g}(t) = \operatorname*{ess\,sup}_{t \in [\xi^j,T]} \pi_{h+\alpha' g}(t) \qquad \forall (h+\alpha' g) \in H.$$

Hence if additionally $\sigma_h^j = \operatorname*{ess\,sup}_{t \in [0,T]} \pi_h^j(t)$, then by (3.2.32b), (3.2.35c) and (7.1.23)

$$\frac{1}{\alpha'}(\sigma_{h+\alpha' g}^j - \sigma_h^j) \geqslant \frac{1}{h}\left[\operatorname*{ess\,sup}_{t \in [\xi^j,T]} \pi_{h+\alpha' g}^j(t) - \operatorname*{ess\,sup}_{t \in [\xi^j,T]} \pi_h^j(t)\right] =$$

$$= \operatorname*{ess\,sup}_{t \in [\xi^j,T]} \frac{1}{h}\left[\pi_{h+\alpha' g}^j(t) - \pi_h^j(t)\right],$$

which together with (7.1.17) imply (7.1.21c).

To prove the complementary slackness (7.1.22) we have to show that the conditions analogous to (3.2.35a) and (3.2.35c) hold. Namely that

$\rho^j(.)=\text{const}$ a.e. in a neighbourhood of t if

$t\in\Xi^j(h)$ or if $t\in\Xi^j\setminus\Xi^j_c$ and

$$\langle D_x\theta^j(x_h(t),h),y(t)\rangle+\langle D_h\theta^j(x_h(t),h),g\rangle<0\,, \tag{7.1.24}$$

$$\zeta^j=\underset{t\in[\xi^j,T]}{\text{ess sup}}\ \rho(t) \quad \text{if} \quad \langle D_x\theta^j(x_h(T),h),y(T)\rangle+\langle D_h\theta^j(x_h(T),h),g)<0. \tag{7.1.25}$$

To show (7.1.24) let us consider the sets:

(6) $(0,T)\setminus\Xi^j(h)$.

By the definition (7.1.5a) we have $\theta^j(x_h(t),h)<0$ for $t\in(0,T)\setminus\Xi^j(h)$. Let us denote $\theta^j(x_h(t),h)=-\gamma<0$. The condition (C5) and (7.1.16) imply that there exists $\bar{\alpha}>0$ such that $\theta^j(x_{h+\alpha'g}(t),h+\alpha'g)\leqslant-\frac{\gamma}{2}$ for any $\alpha'<\bar{\alpha}$. By (C5) and by Lipschitz continuity of x_h, we find that there exists an open subinterval $M\subset(0,T)\setminus\Xi^j(h)$ such that $t\in M$ and $\theta^j(x_{h+\alpha'g}(\tau),h+\alpha'g)\leqslant-\frac{\gamma}{4}$ for any $\alpha'\in[0,\bar{\alpha}]$ and any $\tau\in M$. Hence by (3.2.35a) for any $\alpha'\in[0,\bar{\alpha}]$ we have $\pi_{h+\alpha'g}(\cdot)=\text{const}$ on M, and by (7.1.17e) $\rho(\cdot)=\text{const}$. a.e. on M.

(7) $\{t\in\Xi^j\setminus\Xi^j_c \mid \langle D_x\theta^j(x_h(t),h),y(t)\rangle+\langle D_h\theta^j(x_h(t),h),g\rangle<0\}$

Let $t\in\Xi^j\setminus\Xi^j_c$ and $\langle D_x\theta^j(x_h(t),h)y(t)\rangle+\langle D_h\theta^j(x_h(t),h),g\rangle=\alpha<0$. Hence identically as in (3) and (5) we prove that there exists an open subinterval $M\ni t$ such that $\pi_{h+\alpha'g}(\cdot)=\text{const}$ on M for any $\alpha'\geqslant0$ sufficiently small. This fact together with (7.1.17e) imply $\rho(\cdot)=\text{const}$ a.e. on M.

By (6) and (7) we obtain (7.1.25).

Now, let us prove (7.1.25). If

$$\langle D_x\theta^j(x_h(T),h),y(T)\rangle+\langle D_h\theta^j(x_h(T),h),g\rangle<0, \tag{7.1.26}$$

then by (7.1.14a) $T\notin\Xi^j_c$ and (7.1.23) holds. On the other hand in exactly the same way as in (3) it can be shown that if (7.1.26) holds, then we can choose $\xi^j<T$ such that $\theta^j(x_{h+\alpha'g}(\tau),h+\alpha'g)<0$ for any $\tau\in[\xi^j,T]$ and for any $\alpha'>0$ sufficiently small. Hence by (3.2.35c) for $\alpha'>0$ sufficiently small

$$\sigma^j_{h+\alpha'g}=\underset{t\in[0,T]}{\text{ess sup}}\ \pi^j_{h+\alpha'g}(t)=\pi^j_{h+\alpha'g}(\tau) \quad \text{for } \tau\in[\xi^j,T]. \tag{7.1.27a}$$

By (7.1.27a) and (7.1.16) we get

$$\sigma_h^j = \operatorname*{ess\ sup}_{t\in[0,T]} \pi_h^j(t) = \pi_h^j(\tau) \quad \text{for} \quad \tau \in [\xi^j, T]. \tag{7.1.27b}$$

The equalities (7.1.27) together with (7.1.17) and (7.1.21a) imply (7.1.25).

Now we shall prove that (7.1.24) and (7.1.25) really imply (7.1.22). Let us consider a closed set

$$\Delta^j(h) = \{t \in \Xi^j(h) \mid <D_x\theta^j(x_h(t),h), y(t)> + <D_h\theta^j(x_h(t),h), g> = 0\}.$$

We have

$$\frac{d}{dt}(<D_x\theta^j(x_h(t),h), y(t)> + <D_h\theta^j(x_h(t),h), g>) = 0 \quad \text{a.e. on} \quad \Delta^j(h),$$

hence

$$\int_{\Delta^j(h)} \rho^j(t)\cdot \frac{d}{dt}(<D_x\theta^j(x_h(t),h) y(t)> + <D_h\theta^j(x_h(t),h), g>)dt = 0 \tag{7.1.28}$$

and it is enough to consider the open set $\Gamma^j(h) = (0,T) \setminus \Delta^j(h)$. It consists of at most a countable number of disjoint intervals

$$\Gamma_\eta^j(h) = (\hat{t}_\eta^j, \check{t}_\eta^j), \text{ where } \quad \hat{t}_1^j = 0\ , \quad \check{t}_\eta^j \leqslant T.$$

It follows from (7.1.24) that

$$\rho^j(t) = \rho_\eta^j = \text{const} \quad \text{for a.a.} \quad t \in \Gamma_\eta^j(h). \tag{7.1.29}$$

Note that for all subintervals $\Gamma_\eta^j(h)$, with exception of the first and possibly the last one, we have

$$<D_x\theta^j(x_h(\hat{t}_\eta^j),h); y_h(\hat{t}_\eta^j)> + <D_h\theta^j(x_h(\hat{t}_\eta^j),h), g> =$$

$$= <D_x\theta^j(x_h(\check{t}_\eta^j),h), y_h(\check{t}_\eta^j)> + <D_h\theta^j(x_h(\check{t}_\eta^j),h), g> = 0. \tag{7.1.30}$$

Taking into account (7.1.21b), (7.1.29) and (7.1.30) we get

$$-\sum_\eta \int_{\Gamma_\eta^j} \rho_\eta^j(t)\frac{d}{dt}[<D_x\theta^j(x_h(t),h), y_h(t)> + <D_h\theta^j(x_h(t),h) g>]dt =$$

$$= \sum_\eta \rho_\eta^j\{[<D_x\theta^j(x_h(\check{t}_\eta^j),h), y(\check{t}_\eta^j)> + <D_h\theta^j(x_h(\check{t}_\eta^j),h), g>] +$$

$$- [<D_x\theta^j(x_h(\hat{t}_\eta^j),h), y(\hat{t}_\eta^j)> + <D_h\theta^j(x_h(\hat{t}_\eta^j),h), g>] =$$

$$= \begin{cases} 0 \quad \text{if} \quad [\langle D_x\theta^j(x_h(T),h),y(T)\rangle + \langle D_h\theta^j(x_h(T),h),g\rangle] = 0 \\ -\rho^j(T)[\langle D_x\theta^j(x_h(T),h),y(T)\rangle + \langle D_h\theta^j(x_h(T),h),g\rangle] \quad \text{if} \\ \qquad [\langle D_x\theta^j(x_h(T),h),y(T)\rangle + \langle D_h\theta^j(x_h(T),h),g\rangle] < 0 . \end{cases} \tag{7.1.31}$$

Note that by (7.1.25)

$$\zeta^j = \rho^j(T) \quad \text{if} \quad [\langle D_x\theta^j(x_h(T),h),y(T)\rangle + \langle D_h\theta^j(x_h(T),h),g\rangle] < 0 . \tag{7.1.32}$$

From (7.1.28), (7.1.31) and (7.1.32) it follows that (7.1.22) holds. This concludes the proof of Theorem 7.1

□

Remark 7.1

We are not able to prove Theorem 7.1 without the assumption (C9), since in the case where meas $(\Sigma^j \setminus \Sigma_c^j) > 0$ we fail to show for $(QC_{h,g})$ the complementary slackness condition analogous to (3.2.30). The difficulty arrises from the fact that both in (7.1.7a) and in (7.1.7d) we have only weak convergence in $L^2(0,T)$.

7.2. Continuous Differentiability

Using the same argument as in the proof of Theorem 7.1 we find that the left-differentials

$$v_{h,g}^- = \delta_h^-(h,g) \qquad \text{and} \qquad y_{h,g}^- = \delta_h^-(h,g)$$

of u_h and x_h at h in the direction g are given as the solution of the following quadratic optimal control problem

$$(QC_{h,g}^-) \quad \left| \begin{array}{l} \text{find } (v_{h,g}^-, y_{h,g}^-) \in L^2(0,T) \times W^{1,1}(0,T) \quad \text{such that} \\ K(v_{h,g}, y_{h,g}; h,g) = \min K(v,y;h,g), \\ \text{subject to} \\ \dot{y}(t) = A(h)y(t) + B(h)v(t) + (D_hA(h)g)x_h(t) + (D_hB(h)g)u_h(t), \\ y(0) = 0, \\ \text{and} \\ v(t) \in V_{h,g}^{ad}(t), \\ y(t) \in Y_{h,g}^{ad^-}(t), \end{array} \right.$$

where

$$Y^{ad^-}_{h,g}(t)=\{y\in R^{\ell}\mid \langle D_x\theta^j(x_h(t),h)y\rangle+\langle D_h\theta^j(x_h(t),h),g\rangle \begin{cases}=0 & \text{if } t\in\Xi^j_c(h)\\ \geqslant 0 & \text{if } t\in\Xi^j(h)\setminus\Xi^j_c(h)\end{cases}\}$$

and all other terms are the same as in $(QC_{h,g})$.

It is obvious that in general the solutions to $(QC_{h,g})$ and to $(QC^-_{h,g})$ are different. However in the case where

$$\text{meas }\{t\in\Xi^j(h)\setminus\Xi^j_c(h)\}=0 \tag{7.2.1}$$

the solutions to $(QC_{h,g})$ and to $(QC^-_{h,g})$, as well as the associated Lagrange multipliers, coincide for any direction $g\in R^m$.

Hence we obtain:

Proposition 7.1

If the conditions (C1) through (C9) and (7.2.1) hold, then the solutions (u_h,y_h) of (OC_h) and the associated Lagrange multipliers $(p_h,\lambda_h,\pi_h,\sigma_h)$ are continuously Gâteaux differentiable at h.

As in Section 6.2 also in the case of (OC_h) the condition (7.2.1) is not needed to receiving continuous differentiability of the optimal value function

$$F^o(\cdot):H\to R^1,$$

$$F^o(h)\overset{\text{def}}{=}F(u_h,x_h,h). \tag{7.2.2}$$

Indeed, by (3.1.2), (3.2.30) and (3.2.31) the last four terms in the Lagrangian (3.2.26) vanish at $(u_h,x_h;p_h,\lambda_h,\pi_h,\sigma_h;h)$ and we obtain

$$F^o(h)=L_2(u_h,x_h;p_h,\lambda_h,\pi_h,\sigma_h;h). \tag{7.2.3}$$

Hence

$$\delta^+_{h,g}F^o(h)=(D_uL_2(u_h,x_h;p_h,\lambda_h,\pi_h,\sigma_h;h),v_{h,g})+(D_xL_2(u_h,x_h;p_h,\lambda_h,\pi_h,\sigma_h;h),y_{h,g})+$$

$$+(D_pL_2(u_h,x_h;p_h,\lambda_h,\pi_h,\sigma_h;h),r_{h,g})+(D_\lambda L_2(u_h,x_h;p_h,\lambda_h,\pi_h,\sigma_h;h),\mu_{h,g})+$$

$$+(D_\pi L_2(u_h,x_h;p_h,\lambda_h,\pi_h,\sigma_h;h),\rho_{h,g})+\langle D_\sigma L_2(u_h,x_h;p_h,\lambda_h,\pi_h,\sigma_h;h),\zeta_{h,g}\rangle+$$

$$+\langle D_hL_2(u_h,x_h;p_h,\lambda_h,\pi_h,\sigma_h;h),g\rangle. \tag{7.2.4}$$

Note that (3.2.29) and (3.2.28) imply respectively

$$(D_uL_2(u_h,x_h;p_h,\lambda_h,\pi_h,\sigma_h;h),v_{h,g})=0 \tag{7.2.5}$$

and

$$(D_x L_2(u_h,x_h;p_h,\lambda_h,\pi_h,\sigma_h;h), y_{h,g}) = 0. \tag{7.2.6}$$

On the other hand by (7.1.5a) and (7.1.15a)

$$(D_\lambda L_2(u_h,x_h;p_h,\lambda_h,\pi_h,\sigma_h;h),\mu_{h,g}) = 0. \tag{7.2.7}$$

Taking into account (7.1.6a), (7.1.15b) and (7.1.25) and using the same argument as in the proof of (7.1.22) we get

$$(D_\pi L_2(u_h,x_h;p_h,\lambda_h,\pi_h,\sigma_h;h),\rho_{h,g})+\langle D_\sigma L_2(u_h,x_h;p_h,\lambda_h,\pi_h,\sigma_h;h),\zeta_{h,g}\rangle = 0. \tag{7.2.8}$$

Substituting (7.2.5)-(7.2.8) into (7.2.4) we obtain

$$\delta^+_{h,g}F^0(h) = \langle D_h L_2(u_h,x_h;p_h,\lambda_h,\pi_h,\sigma_h;h),g\rangle. \tag{7.2.9a}$$

Similarly

$$\delta^-_{h,g}F^0(h) = \langle D_h L_2(u_h,x_h;p_h,\lambda_h,\pi_h,\sigma_h;h),g\rangle. \tag{7.2.9b}$$

Note that by (C2), (C3), (C5) as well as by Theorem 3.2 and Corollary 3.8

$$D_h L_2(u_h,x_h;p_h,\lambda_h,\pi_h,\sigma_h;h) = D_h F(u_h,x_h,h) - (p_h, D_h A(h)x_h + D_h B(h)u_h) +$$

$$+ (\lambda_h, D_h\phi(u_h,h)) - (\pi_h, D^2_{xh}\theta(x_h,h)\dot{x}_h) + \langle\sigma_h, D_h\theta(x_h(T),h)\rangle \tag{7.2.10}$$

is a continuous function of h.

Hence from (7.2.9) we obtain

<u>Proposition 7.2</u>

If the conditions (C1) through (C9) hold, then the optimal value function $F^0(\cdot)$ for (OC_h) is continuously (Fréchet) differentiable at any $h \in H$ and

$$D_h F^0(h) = D_h L_2(u_h,x_h;p_h,\lambda_h,\pi_h,\sigma_h;h). \tag{7.2.11}$$

Note that the result (7.2.11) is well known in stability of optimal control problems in a much more general case (see [20, 42]).

CONCLUDING REMARKS

The result presented in these notes concern Lipschitz continuity and directional differentiability of strongly convex mathematical programming and optimal control problems that depend on a vector parameter.

In deriving these results the Lagrange formalism for initial optimization problems is used.

The results are fairly complete for mathematical programming problems. The method is applicable to a broad class of optimal control problems subject to control constraints with lumped or distributed parameters, provided that control constraints are of pointwise type. Also additional unilateral constraints in the form of a finite number of convex functionals can be easily handled.

As it can be seen in Chapter 3 and 7 the optimal control problems with state space constraints create much more serious difficulties. The considered case of a system described by ordinary differential equations requires a very elaborate analysis and the results concerning differential stability are far from being complete.

Distributed parameter systems subject to state space constraints are very hard to be attacked using the presented method, since the Lagrange formalism for these systems is not well developed.

The methodology used has some shortcommings:

1) it requires uniquuess not only of the solutions, but also of the associated Lagrange multipliers, which narrows down the scope of possible applications,
2) it seems that the method can not be applied to abstract convex problems of optimization in functional spaces.

Maybe these difficulties can be overcome using the method of projection onto convex sets in Hilbert spaces [27,43,48], which does not involve Lagrange formalism, and which, at least in the case of linear constraints, gives good results. However up to now the forms of the directional differential of the operation of projection are not known in more general cases.

To conclude we formulate several unsolved problems directly connected with the presented material:

- problem of higher order directional differentiability in arbitrary directions of solutions to convex programms (see Remark 4.1) and higher order directional differentiability of solutions to optimal control problems,
- full charcterization of Clarke's generalized gradients, with respect to the parameter, of solutions to convex programming (see Section 4.5)

and optimal control problems,

- proof of Theorem 7.1 without assumption (C9). In the presented proof this assumption seems to be necessary due rather to technical difficulties, than to the very nature of the problem,
- problem of strong convergence of difference quotients of optimal controls and Lagrange multipliers for control problems subject to state space constraints (see(7.1.7)).

REFERENCES

[1] R.A. ADAMS, Sobolev Spaces, Academic Press, New York, 1975.

[2] A. AUSLENDER, Differential Stability in Nonconvex and Nondifferentiable Programming, in: Mathematical Programming Study, vol. 10, P. Huard (ed.), North-Holland, Amsterdam 1979, pp. 29-41.

[3] A. AUSLENDER, Stability in Mathematical Programming with Nondifferentiable Data, SIAM J. Control Optim. 22 (1984), 239-254.

[4] B. BANK, J. GUDDAT, D. KLATTE, B. KUMMER, K. TAMMER, Nonlinear Parametric Optimization, Academie-Verlag, Berlin 1982.

[5] H. BIGELOW, N.Z. SHAPIRO, Implicit Function Theorems for Mathematical Programming and for Systems of Inequalities, Mathematical Programming, 6 (1974), 141-156.

[6] M.D. CANON, C.D. CULLUM, E. POLAK, Theory of Optimal Control and Mathematical Programming, Mc Graw-Hill, New York, 1970.

[7] F.H. CLARKE, Generalized Gradients and Applications, Trans. Am. Math. Soc. 205 (1975), 247-262.

[8] F.H. CLARKE, A New Approach to Lagrange Multipliers, Math. Operations Res. 1 (1976), 165-174.

[9] F.H. CLARKE, Optimization and Nonsmooth Analysis, J. Wiley and Sons, New York, 1983.

[10] B. CORNET, J.-Ph. VIAL, Lipschitzian Solutions of Perturbed Nonlinear Programming Problems, CORE Discussion Paper, No. 8311, Université Catholique de Louvain, Louvain-la-Neuve, 1983.

[11] R.S. DEMBO, Sensitivity Analysis in Geometric Programming, J. Optim. Theory Appl. 37 (1982), 1-21.

[12] A.L. DONTCHEV, Perturbations, Approximations and Sensitivity Analysis of Optimal Control Systems. Lecture Notes in Control and Information Sciences, Vol 52, Springer-Verlag, Berlin, 1983.

[13] A.L. DONTCHEV, H.TH. JONGEN, On the Regularity of the Kuhn-Tucker Curve, Memorandum Nr. 460, Department of Applied Mathematics, Twente University of Technology, Enschede, 1984.

[14] N. DUNFORD, J.T. SCHWARZ, Linear Operators, Part I, Inter-Science Publischers Inc., New York, 1958.

[15] A.V. FIACCO, Sensitivity Analysis for Nonlinear Programming Using Penalty Methods, Math. Programming 10 (1976), 287-311.

[16] A.V. FIACCO, Nonlinear Programming Sensitivity Analysis Results Using Strong Second Order Assumptions, in: Numerical Optimization of Dynamic Systems, L.C.W. Dixon, G.P. Szegö (eds.), North Holland, Amsterdam, 1980.

[17] A.V. FIACCO, Introduction to Sensitivity and Stability Analysis

in Nonlinear Programming, Academic Press, New York, London, 1983.

[18] J. GAUVIN, The Generalized Gradient of the Marginal Function in Mathematical Programming, Math. Operations Res. 4 (1979), 458-463.

[19] B. GOLLAN, A General Perturbation Theory for Abstract Optimization Problems, J. Opt. Theory Appl. 35 (1981), 417-441.

[20] B. GOLLAN, On the Optimal Value Function of Optimal Control Problems, Zeitschrift fur Analysis und ihre Anwendingen 1 (1982), 17-33.

[21] B. GOLLAN, On the Marginal Function in Nonlinear Programming, Math. Operations Res. 9 (1984), 208-221.

[22] E.G. GOL'SHTEIN, Duality Theory in Mathematical Programming and Its Applications, Nauka, Moscow 1971 (in Russian).

[23] T.H. GRONWALL, Note on the Derivatives with Respect to a Parameter of the Solutions of a System of Differential Equations, Ann. Math. 20 (1919), 292-296.

[24] W.W. HAGER, Lipschitz Continuity for Constrained Processes, SIAM J. Control Optim. 17 (1979), 321-333.

[25] W.W. HAGER, S.K. MITTER, Lagrange Duality Theory for Convex Control Problems, SIAM J. Control Optim. 14 (1976), 843-856.

[26] A. HARAUX, Derivation dans les Inéquations Variationelles, C.R. Acad. Sci., Serie A, 278 (1974), 1257-1260.

[27] A. HARAUX, How to Differentiate the Projection on a Convex Set in Hilbert Space. Some Application to Variational Inequalities, J. Math. Soc. Japan, 29 (1977), 615-631.

[28] M. R. HESTENES, Calculus of Variations and Optimal Control Theory, Wiley, New York, 1966.

[29] J.B. HIRIART-URRUTY, At What Points is the Projection Mapping Differentiable?, Amer. Math. Monthly 89 (1982), 456-458.

[30] K. JITTORNTRUM, Sequential Algorithms in Nonlinear Programming, Ph. D. dissertation, Australian National Univ., Canberra, 1978.

[31] K. JITTORNTRUM, Solution Point Differentiability without Strict Complementarity in Nonlinear Programming, in: Mathematical Programming Study, Vol. 21, A.V. Fiacco (ed.), North-Holland, Amsterdam, 1984, pp. 127-138.

[32] M. KOJIMA, Strongly Stable Stationary Solutions in Nonlinear Programs, in: Analysis and Computation of Fixed Points, S.M. Robinson (ed.), Academic Press, New York, 1980, pp. 93-138.

[33] E.B. LEE, L. MARKUS, Foundations of Optimal Control Theory, J. Wiley, New York, 1967.

[34] F. LEMPIO, H. MAURER, Differential Stability in Infinite-Dimensional Nonlinear Programming, Appl. Math. Optimiz. 6 (1980), 139-152.

[35] J.L. LIONS, Contrôl Optimal des Systèmes Gonvernés par des Équations aux Dérivées Partielles, Dunod, Gauther-Villars, Paris, 1968.

[36] J.L. LIONS, E. MAGENES, Problèmes aux Limites Non Homogènes et Applications, Vol. 1, 2, Dunod, Paris, 1968.

[37] K. MALANOWSKI, Differential Stability of Solutions to Convex Control Constrained Optimal Control Problems, Appl. Math. Optimiz. 12 (1984), 1-14.

[38] K. MALANOWSKI, On Differentiability with Respect to Parameter of Solutions to Convex Optimal Control Problems Subject to State Space Constraints, Appl. Math. Optimiz. 12 (1984), 231-245.

[39] K. MALANOWSKI, Differentiability with Respect to Parameters of Solutions to Convex Programming Problems, Math.Progr. 33 (1985) 352-361.

[40] K. MALANOWSKI, Differential Sensitivity of Solutions to Convex Constrained Optimal Control Problems for Discrete Systems (to be published).

[41] K. MALANOWSKI, J. SOKOŁOWSKI, Sensitivity of Solutions to Convex Control Constrained Optimal Control Problems for Distributed Parameter Systems (to be published).

[42] H. MAURER, Differential Stability in Optimal Control Problems, Appl. Math. Optimiz. 5 (1979), 63-82.

[43] F. MIGNOT, Contrôl dans les Inéquations Variationelles, J. Functional Anal., 22, (1976), 130-185.

[44] S.M. ROBINSON, Perturbed Kuhn-Tucker Points and Rate of Convergence for a Class of Nonlinear Programming Algorithms, Math. Programming 7 (1974), 1-16.

[45] S.M. ROBINSON, Strongly Regular Generalized Equations, Math. Operations Res., 5 (1) (1980), 43-62.

[46] R.T. ROCKAFELLAR, Lagrange Multipliers and Subderivatives of Optimal Value Functions in Nonlinear Programming, in: Mathematical Programming Study, Vol. 17, D.C. Sorensen and R.J.-B. Wets (eds.), North-Holland, Amsterdam 1982, pp. 28-62.

[47] R.T. ROCKAFELLAR, Directional Differentiability of the Optimal Value Function in a Nonlinear Programming Problem, in: Mathematical Programming Study, Vol. 21, A.V. Fiacco (ed.), North-Holland, Amsterdam 1984, pp. 213-226.

[48] J. SOKOŁOWSKI, Differential Stability of Solutions to Constrained Optimization Problems, Appl. Math. Optimiz. 13 (1985), 97-115.

[49] J. SOKOŁOWSKI, Sensitivity Analysis of Control Constrained Optimal Control Problems for Distributed Parameter Systems, (to be

published).

[50] J. SOKOŁOWSKI, Differential Stability of Control Constrained Optimal Control Problems for Distributed Parameter Systems, in: Proceedings of the 2nd International Conference on Control Theory for Distributed Parameter Systems and Applications, Vorau, 1984 (to appear).

[51] J. SOKOŁOWSKI, Problems of Sensitivity Analysis and Parametric Optimization in Optimal Control of Distributed Parameter Systems, dissertation, Technical University of Warsaw, Zeszyty Naukowe. Elektronika.z.73. Wydawnictwa Politechniki Warszawskiej,Warszawa,1985 (in Polish).

[52] H. TRIEBEL, Interpolation Theory, Function Spaces, Differential Operators, VEB Deutscher Verlag der Wissenschaften, Berlin, 1978.

Lecture Notes in Control and Information Sciences

Edited by M. Thoma

Vol. 43: Stochastic Differential Systems
Proceedings of the 2nd Bad Honnef Conference
of the SFB 72 of the DFG at the University of Bonn
June 28 – July 2, 1982
Edited by M. Kohlmann and N. Christopeit
XII, 377 pages. 1982.

Vol. 44: Analysis and Optimization of Systems
Proceedings of the Fifth International
Conference on Analysis and Optimization of Systems
Versailles. December 14–17, 1982
Edited by A. Bensoussan and J. L. Lions
XV, 987 pages, 1982

Vol. 45: M. Arató
Linear Stochastic Systems
with Constant Coefficients
A Statistical Approach
IX, 309 pages. 1982

Vol. 46: Time-Scale Modeling of Dynamic Networks
with Applications to Power Systems
Edited by J. H. Chow
X, 218 pages. 1982

Vol. 47: P. A. Ioannou, P. V. Kokotovic
Adaptive Systems with Reduced Models
V, 162 pages. 1983

Vol. 48: Yaakov Yavin
Feedback Strategies for Partially
Observable Stochastic Systems
VI, 233 pages, 1983

Vol. 49: Theory and Application of Random Fields
Proceedings of the IFIP-WG 7/1
Working Conference
held under the joint auspices of the
Indian Statistical Institute
Bangalore, India, January 1982
Edited by G. Kallianpur
VI. 290 pages. 1983

Vol. 50: M. Papageorgiou
Applications of Automatic Control Concepts
to Traffic Flow Modeling and Control
IX, 186 pages. 1983

Vol. 51: Z. Nahorski, H.F. Ravn, R.V.V. Vidal
Optimization of Discrete Time Systems
The Upper Boundary Approach
V, 137 pages 1983

Vol. 52: A. L. Dontchev
Perturbations, Approximations and Sensitivity Analysis
of Optimal Control Systems
IV, 158 pages. 1983

Vol. 53: Liu Chen Hui
General Decoupling Theory of Multivariable
Process Control Systems
XI, 474 pages. 1983

Vol. 54: Control Theory for Distributed
Parameter Systems and Applications
Edited by F. Kappel, K. Kunisch,
W. Schappacher
VII, 245 pages. 1983.

Vol. 55: Ganti Prasada Rao
Piecewise Constant Orthogonal Functions
and Their Application to Systems and Control
VII, 254 pages. 1983.

Vol. 56: Dines Chandra Saha, Ganti Prasada Rao
Identification of Continuous
Dynamical Systems
The Poisson Moment Functional
(PMF) Approach
IX, 158 pages. 1983.

Vol. 57: T. Söderström, P. G. Stoica
Instrumental Variable Methods
for System Identification
VII, 243 pages. 1983.

Vol. 58: Mathematical Theory of
Networks and Systems
Proceedings of the MTNS-83 International
Symposium
Beer Sheva, Israel, June 20–24, 1983
Edited by P. A. Fuhrmann
X, 906 pages. 1984

Vol. 59: System Modelling and Optimization
Proceedings of the 11th IFIP Conference
Copenhagen, Denmark, July 25-29, 1983
Edited by P. Thoft-Christensen
IX, 892 pages. 1984

Vol. 60: Modelling and Performance
Evaluation Methodology
Proceedings of the International Seminar
Paris, France, January 24–26, 1983
Edited by F. Bacelli and G. Fayolle
VII, 655 pages. 1984

Vol. 61: Filtering and Control of Random
Processes
Proceedings of the E.N.S.T.-C.N.E.T. Colloquium
Paris, France, February 23–24, 1983
Edited by H. Korezlioglu, G. Mazziotto, and
J. Szpirglas
V, 325 pages. 1984

Lecture Notes in Control and Information Sciences

Edited by M. Thoma and A. Wyner

Vol. 62: Analysis and Optimization of Systems
Proceedings of the Sixth International Conference on Analysis and Optimization of Systems
Nice, June 19–22, 1984
Edited by A. Bensoussan, J. L. Lions
XIX, 591 pages. 1984.

Vol. 63: Analysis and Optimization of Systems
Proceedings of the Sixth International Conference on Analysis and Optimization of Systems
Nice, June 19–22, 1984
Edited by A. Bensoussan, J. L. Lions
XIX, 700 pages. 1984.

Vol. 64: Arunabha Bagchi
Stackelberg Differential Games in Economic Models
VIII, 203 pages, 1984

Vol. 65: Yaakov Yavin
Numerical Studies in Nonlinear Filtering
VIII, 273 pages, 1985.

Vol. 66: Systems and Optimization
Proceedings of the Twente Workshop
Enschede, The Netherlands, April 16–18, 1984
Edited by A. Bagchi, H. Th. Jongen
X, 206 pages, 1985.

Vol. 67: Real Time Control of Large Scale Systems
Proceedings of the First European Workshop
University of Patras, Greece, Juli 9–12, 1984
Edited by G. Schmidt, M. Singh, A. Titli, S. Tzafestas
XI, 650 pages, 1985.

Vol. 68: T. Kaczorek
Two-Dimensional Linear Systems
IX, 397 pages, 1985.

Vol. 69: Stochastic Differential Systems – Filtering and Control
Proceedings of the IFIP-WG 7/1 Working Conference
Marseille-Luminy, France, March 12-17, 1984
Edited by M. Metivier, E. Pardoux
X, 310 pages, 1985.

Vol. 70: Uncertainty and Control
Proceedings of a DFVLR International Colloquium
Bonn, Germany, March, 1985
Edited by J. Ackermann
IV, 236 pages, 1985.

Vol. 71: N. Baba
New Topics in Learning Automata Theory and Applications
VII, 231 pages, 1985.

Vol. 72: A. Isidori
Nonlinear Control Systems: An Introduction
VI, 297 pages, 1985.

Vol. 73: J. Zarzycki
Nonlinear Prediction Ladder-Filters for Higher-Order Stochastic Sequences
V, 132 pages, 1985.

Vol. 74: K. Ichikawa
Control System Design based on Exact Model Matching Techniques
VII, 129 pages, 1985.

Vol. 75: Distributed Parameter Systems
Proceedings of the 2nd International Conference, Vorau, Austria 1984
Edited by F. Kappel, K. Kunisch, W. Schappacher
VIII, 460 pages, 1985.

Vol. 76: Stochastic Programming
Edited by F. Archetti, G. Di Pillo, M. Lucertini
V, 285 pages, 1986.

Vol. 77: Detection of Abrupt Changes in Signals and Dynamical Systems
Edited by M. Basseville, A. Benveniste
X, 373 pages, 1986.

Vol. 78: Stochastic Differential Systems
Proceedings of the 3rd Bad Honnef Conference, June 3–7, 1985
Edited by N. Christopeit, K. Helmes, M. Kohlmann
V, 372 pages, 1986.

Vol. 79: Signal Processing for Control
Edited by K. Godfrey, P. Jones
XVIII, 413 pages, 1986.

Vol. 80: Artificial Intelligence and Man-Machine Systems
Edited by H. Winter
IV, 211 pages, 1986.

Lecture Notes in Control and Information Sciences

Edited by M. Thoma and A. Wyner